my **revision** notes

Pearson Edexcel

INTERNATIONAL GCSE (9-1) MATHEMATICS FOR SPECIFICATION A

Sophie Goldie
Sadhiv Mahandru

Series editor:
Jean Linsky

The Publishers would like to thank the following for permission to reproduce copyright material.

Acknowledgements

Every effort has been made to trace all copyright holders, but if any have been inadvertently overlooked, the Publishers will be pleased to make the necessary arrangements at the first opportunity.

Orders: please contact Bookpoint Ltd, 130 Park Drive, Milton Park, Abingdon, Oxon OX14 4SE. Telephone: (44) 01235 827827. Fax: (44) 01235 400401. Email education@bookpoint.co.uk. Lines are open from 9 a.m. to 5 p.m., Monday to Saturday, with a 24-hour message answering service. You can also order through our website: www.hoddereducation.co.uk

ISBN: 978 1 5104 4692 2

First published in 2019 by

Hodder Education,
An Hachette UK Company
Carmelite House
50 Victoria Embankment
London EC4Y 0DZ

www.hoddereducation.co.uk

Impression number 10 9 8 7 6 5 4 3 2 1

Year 2023 2022 2021 2020 2019

Typeset in Integra Software Services Pvt. Ltd., Pondicherry, India

Printed in Spain

A catalogue record for this title is available from the British Library.

Get the most from this book

Welcome to your Mathematics Revision Guide for the International GCSE for Pearson Edexcel. This book will provide you with reminders of the knowledge and skills you will be expected to demonstrate in the exam with opportunities to check and practice those skills on exam-style questions. Additional hints and notes throughout help you to avoid common errors and provide a better understanding of what's needed in the exam.

Included with the purchase of this book is valuable online material that provides full worked solutions to all the 'Target your revision', 'Exam-style questions' and 'Review questions'. The **online material** is available at www.hoddereducation.co.uk/MRNEdexIGCSEMaths

Features to help you succeed

Target your revision

Use these questions at the start of each of the six sections to focus your revision on the topics you find tricky. **Short answers** are at the back of the book, but use the **worked solutions online** to check each step in your solution.

Key facts

Check you understand all the key facts in each subsection. These provide a useful checklist if you get stuck on a question.

Worked examples

Full worked examples show you what the examiner expects to see in order to ensure full marks in the exam. The examples cover a wide range of the type of questions you can expect.

Remember

Handy reminders to keep you on track.

Watch out!

Common pitfalls to be aware of to avoid slip-ups.

Exam tip

Helpful pointers to maximise your marks in the exam.

Exam-style questions

For each topic, these provide typical questions you should expect to meet in the exam. **Short answers** are at the back of the book, and you can check your working using the **online worked solutions**.

Review questions

After you have completed each of the six sections in the book, answer these questions for more practice. **Short answers** are supplied at the back of the book, and there are **full worked solutions online** to allow you to check every line in your solution.

At the end of the book, you will find some useful information:

Revision tips

Includes hints and tips on revising for the International GCSE exam, and details the exact structure of the exam papers.

Make sure you know these formulae for your exam

Provides a succinct list of all the formulae you need to remember and the formulae that will be given to you in the exam.

Please note that the formula sheet as provided by the exam board for the exam may be subject to change.

During your exam

Includes key words to watch out for, common mistakes to avoid and tips if you get stuck on a question.

My revision planner

REVISED | TESTED | EXAM READY

SECTION 5

SECTION 6

Go online to www.hoddereducation.co.uk/MRNEdexIGCSEMaths for:

- full worked solutions to the Target your revision questions
- full worked solutions to all Exam-style questions
- full worked solutions to all Review questions

Target your revision: Number

Check how well you know each topic by answering these questions. If you struggle, go to the page number in brackets to revise that topic.

1 Add and subtract fractions
Show that

a $1\frac{7}{12} - \frac{5}{6} = \frac{3}{4}$

b $3\frac{7}{15} + 1\frac{5}{6} = 5\frac{3}{10}$

(see page 3)

2 Multiply and divide fractions
Show that

a $2\frac{4}{5} \times 1\frac{5}{6} = 5\frac{2}{15}$

b $3\frac{3}{10} \div 2\frac{2}{5} = 1\frac{3}{8}$

(see page 4)

3 Convert decimals to fractions
Show that

a $0.725 = \frac{29}{40}$

b $0.\dot{1}\dot{8} = \frac{2}{11}$

(see page 5)

4 Solve problems involving ratios
The ratio of width to length of a cargo ship is 3 : 20
The length of the ship is 400 m.

a Work out the width of the ship.

A model of the cargo ship is made.
The length of the model is 25 cm.

b Find the ratio of the length of the model to the length of the real ship.
Give your ratio in the form 1 : n

(see page 6)

5 Solve problems involving proportion
An exchange rate is
1 Indian rupee = 1.63 Japanese yen.
Aki changes 500 yen to rupees.
How many rupees does Aki get?
Give your answer to the nearest rupee.
(see page 7)

6 Solve problems involving percentages
Ella drives to work each day. The weekly cost of her petrol is £36. Ella's weekly pay is £504

a What percentage of Ella's weekly pay does she spend on petrol?

Ella is given a pay rise of 5%.

b What is Ella's weekly pay after her pay rise?

Ella's car cost £16 000 when it was new.
The value of her car has decreased by 45%.

c How much is Ella's car worth now?

(see page 8)

7 Solve compound interest problems
Isobel invested \$4000 for 2 years at 2.5% per annum compound interest.
Calculate the value of her investment at the end of the 2 years.
(see page 9)

8 Solve reverse percentage problems
In a sale, all normal prices are reduced by 35%
The normal price of a jacket is reduced by \$28
Work out the normal price of jacket.
(see page 9)

9 Use the laws of indices

a Write $\frac{2^4 \times 8^2}{2}$ as a single power of 2

b Simplify $\frac{(3a\sqrt{b})^4}{9a^2b}$

(see page 10)

10 Find a product of prime factors
Write 630 as a product of prime factors.
(see page 11)

11 Find HCF and LCM of two numbers
Find the highest common factor (HCF) and the lowest common multiple (LCM) of 210 and 600
(see page 11)

12 Understand standard form

a Write the following as ordinary numbers.
i 7.92×10^6
ii 8.1×10^{-3}

b Arrange these numbers in order of size.
Start with the smallest number.
6.1×10^8 9.5×10^7 9.38×10^7

c Write the following in standard form.
i 89 000 000
ii 0.000 387

(see page 12)

13 Calculate with numbers in standard form

a The volume of Neptune is 6.25×10^{13} km³ and the volume of Uranus is 6.83×10^{13} km³.
Find
i the total volume of Neptune and Uranus,
ii the difference between the volume of Neptune and the volume of Uranus.
Give your answers in standard form.

b The volume of Earth is 1.08×10^{12} km³.
The volume of Neptune is k times larger than the volume of Earth.
Find, to the nearest integer, the value of k.

(see page 12)

14 Round to a given number of significant figures

a Round each of the following to 3 significant figures.
i 2.365
ii 0.020 419
iii 5.997

Number

b Use your calculator to evaluate

$$\frac{\sqrt{30.8}}{2.61-0.74}+7.3^2$$

Give your answer to correct to 3 significant figures.

(see page 13)

15 Find upper and lower bounds

The length of a room is 3.89 metres, correct to the nearest centimetre.

Write down:

i the lower bound for the length of the room,

ii the upper bound for the length of the room.

(see page 13)

16 Understand set notation

$\mathscr{E}$ = {positive integers less than 16}

P = {multiples of 2}

Q = {multiples of 3}

R = {odd numbers}

a List the members of the set:

i $P \cap Q$

ii $Q \cup R$

b Explain why $P \cap R = \varnothing$

c $x \in Q$ and $x \notin P$.

Write down a possible value for x.

d The set S has three elements.

$S \subset Q$ and $S \subset R$

List the members of the set S.

(see page 14)

17 Use Venn diagrams

X and Y are two sets.

$n(\mathscr{E}) = 30$

$n(X) = 10$

$n(Y) = 15$

$n(X \cap Y) = 4$

a Complete the Venn diagram to show the number of elements.

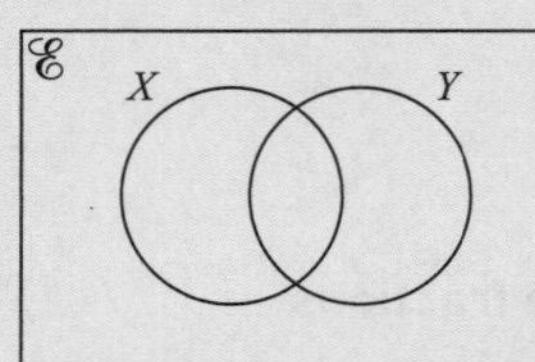

b Find:

i $n(Y')$,

ii $n(X' \cup Y)$.

(see page 14)

18 Simplify surds

Without using a calculator, simplify

i $\sqrt{175}+\sqrt{28}$

ii $\frac{54}{\sqrt{6}}$

iii $(6-2\sqrt{5})(6+2\sqrt{5})$

iv $\frac{7\sqrt{2}}{3-\sqrt{2}}$

(see page 16)

Short answers on page 132

Full worked solutions online

CHECKED ANSWERS ONLINE

Adding and subtracting fractions

REVISED

Number

Key facts

1 Equivalent fractions have the same decimal value.
To find equivalent fractions:

- Multiply or divide the numerator and denominator of the fraction by the same number

Examples: $\frac{2}{7} \overset{\times 3}{=} \frac{6}{21}$ and $\frac{12}{20} \overset{\div 4}{=} \frac{3}{5}$

2 To **add or subtract fractions**:

- Find equivalent fractions with the same denominators
- Add/subtract the numerators
- Simplify your answer

3 A **mixed number** has an integer (whole number) and a fraction part

Example: $3\frac{1}{6} = \frac{18}{6} + \frac{1}{6} = \frac{19}{6}$

Remember

To change a fraction to a decimal, find:
'top' ÷ 'bottom'
So $\frac{3}{5} = 3 \div 5 = 0.6$

Worked examples

Adding fractions

Show that $\frac{3}{8} + \frac{5}{12} = \frac{19}{24}$

Solution

Find a common denominator: $\frac{3\times 3}{8\times 3} + \frac{5\times 2}{12\times 2}$

Add the numerators: $\frac{9}{24} + \frac{10}{24} = \frac{19}{24}$

Using mixed numbers

Show that $4\frac{3}{10} - 2\frac{7}{15} = 1\frac{5}{6}$

Solution

Rewrite as improper fractions: $4\frac{3}{10} = \frac{40}{10} + \frac{3}{10} = \frac{43}{10}$

and $2\frac{7}{15} = \frac{30}{15} + \frac{7}{15} = \frac{37}{15}$

Subtract the improper fractions: $\frac{43}{10} - \frac{37}{15}$

Find a common denominator: $\frac{43\times 3}{10\times 3} - \frac{37\times 2}{15\times 2}$

Subtract the numerators and write the answer as a mixed number:

$\frac{129}{30} - \frac{74}{30} = \frac{55}{30} = 1\frac{25\div 5}{30\div 5} = 1\frac{5}{6}$

Exam tip

You must make sure you show all your working otherwise you will lose marks.

Watch out!

It is best to rewrite any mixed numbers as improper fractions first.
Don't try to work out $4-2$ and $\frac{3}{10} - \frac{7}{15}$ separately, as $\frac{7}{15}$ is greater than $\frac{3}{10}$

Remember

A fraction is in its simplest form when you can't simplify it any further.

Exam-style question

Show that

a $\frac{7}{12} - \frac{2}{9} = \frac{13}{36}$ [2]

b $\frac{5}{12} + \frac{11}{15} = 1\frac{3}{20}$ [3]

c $5\frac{7}{10} - 2\frac{5}{6} = 2\frac{13}{15}$ [3]

Short answers on page 132

Full worked solutions online

CHECKED ANSWERS ONLINE

Multiplying and dividing fractions

REVISED

Key facts

1 To **multiply fractions**:

- Convert any mixed numbers to improper fractions
- Multiply the numerators (tops)
- Multiply the denominators (bottoms)
- Simplify your answer

Example: $\frac{2}{3}\times\frac{5}{7}=\frac{10}{21}$

2 To **divide by a fraction**:

- Turn the 2nd fraction upside down (this is the **reciprocal**)
- Multiply

Example: $\frac{2}{3}\div\frac{5}{7}=\frac{2}{3}\times\frac{7}{5}=\frac{14}{15}$

Remember

'Of' means multiply.

So $\frac{5}{9}$ of 27 means

$$\frac{5}{9}\times27=\frac{5\times27}{9}$$
$$=5\times3$$
$$=15$$

Remember

Dividing by $\frac{1}{n}$ is the same as multiplying by n.

Worked examples

Multiplying fractions

Show that $1\frac{3}{4}\times\frac{2}{5}=\frac{7}{10}$

Solution

Rewrite as improper fractions: $1\frac{3}{4}=\frac{4}{4}+\frac{3}{4}=\frac{7}{4}$

Multiply: $1\frac{3}{4}\times\frac{2}{5}=\frac{7}{4}\times\frac{2}{5}=\frac{14}{20}$

Simplify: $\frac{14\div2}{20\div2}=\frac{7}{10}$

Dividing fractions

Show that $3\frac{1}{3}\div1\frac{3}{5}=2\frac{1}{12}$

Solution

Rewrite as improper fractions: $3\frac{1}{3}=\frac{9}{3}+\frac{1}{3}=\frac{10}{3}$

and $1\frac{3}{5}=\frac{5}{5}+\frac{3}{5}=\frac{8}{5}$

Turn 2nd fraction upside down: $\frac{10}{3}\div\frac{8}{5}=\frac{10}{3}\times\frac{5}{8}$

Multiply: $\frac{\cancel{10}^{5}}{3}\times\frac{5}{\cancel{8}_{4}}=\frac{25}{12}=2\frac{1}{12}$

Exam tip

You must make sure you show all your working otherwise you will lose marks.

Watch out!

Always rewrite any mixed numbers as improper fractions first.

Remember

You might find it easier to cancel out any common factors before you multiply.

Exam-style question

Show that

a $\frac{2}{3}\div\frac{4}{5}=\frac{5}{6}$ [2]

b $2\frac{5}{6}\times1\frac{4}{5}=5\frac{1}{10}$ [3]

c $1\frac{3}{10}\div2\frac{1}{3}=\frac{39}{70}$ [3]

Short answers on page 132

Full worked solutions online

CHECKED ANSWERS ONLINE

Converting decimals to fractions

REVISED

Number

Key facts

1 A **terminating decimal** is a decimal which ends. It has a finite number of digits.

You can convert a terminating decimal to a fraction with a denominator of 10, 100 or 1000 and so on.

Examples: $0.7 = \frac{7}{10}$, $0.53 = \frac{53}{100}$ and $0.625 = \frac{625 \div 125}{1000 \div 125} = \frac{5}{8}$

2 A **recurring decimal** is a decimal with a pattern of digits which repeat indefinitely.

Examples: 0.222…, 0.135135135… and 0.4181818…

You write dots over the start and finish of the repeating pattern.

Examples: $0.\dot{2}$, $0.\dot{1}3\dot{5}$ and $0.4\dot{1}\dot{8}$

3 Use the **multiply/subtract method** to convert a recurring decimal to a fraction.

- Let x = decimal to be converted.
- Count the number of digits that are being repeated.

Number of repeated digits	1	2	3
Multiply by...	10	100	1000

Multiply both sides by 10, 100 or 1000 accordingly.

- Subtract to find the value of $9x$, $99x$ or $999x$ as an integer.
- Divide to find x as a fraction.

Example:

$x = 0.\dot{2}$

So $10x = 2.\dot{2}$

$10x - x = 2.\dot{2} - 0.\dot{2}$

So $9x = 2$

So $x = \frac{2}{9}$

Worked examples

Converting a recurring decimal to a fraction (1)

Show that the recurring decimal $0.\dot{1}3\dot{5} = \frac{5}{37}$

Solution

Let $x = 0.\dot{1}3\dot{5}$

Multiply by 1000: $1000x = 135.\dot{1}3\dot{5}$

Subtract: $1000x - x = 135.\dot{1}3\dot{5} - 0.\dot{1}3\dot{5}$

$999x = 135$

Find x as a fraction: $x = \frac{135 \div 27}{999 \div 27} = \frac{5}{37}$

Converting a recurring decimal to a fraction (2)

Use algebra to show that the recurring decimal $0.4\dot{1}\dot{8} = \frac{23}{55}$

Solution

Let $x = 0.4\dot{1}\dot{8}$

Multiply by 100: $100x = 41.8181818...$

Subtract: $100x - x = 41.81818... - 0.4181818...$

$99x = 41.4$

Find x as a fraction: $x = \frac{41.4 \times 10}{99 \times 10} = \frac{414 \div 18}{990 \div 18} = \frac{23}{55}$

Remember

All fractions convert to a decimal which either terminates or recurs.

Exam tip

You must show every step in your working for 'show that' style questions.

Watch out!

Take extra care when not all the digits after the decimal point are recurring. When you subtract you will end up with a decimal. You can multiply by 10, 100, or 1000 to remove the decimal.

Exam-style question

Use algebra to show that the recurring decimal $0.\dot{3}8\dot{7} = \frac{43}{111}$ [2]

Short answers on page 132

Full worked solutions online

CHECKED ANSWERS ONLINE

Ratio

REVISED

Key facts

1 Ratios are used to compare quantities.

Example: The ratio of girls to boys in a class is 3 : 2

The order the ratio is written in matters.

Example: The ratio of boys to girls in the class is 2 : 3

2 You can simplify ratios by dividing each 'side' of the ratio by the same number.

Example: $10:15:25 \overset{\div 5}{=} 2:3:5$ (÷5)

3 Scales are often written as a ratio of $1:n$

Example: The scale on a map is 1 : 20 000

So 1 cm on the map represents 20 000 cm in real life.

Remember

A ratio is like a fraction.

$\frac{3}{5}$ of the students in the class are girls and $\frac{2}{5}$ are boys.

Worked examples

Solving problems with ratios

April, Ben and Cate share some money in the ratio 2 : 3 : 6

Cate receives \$30 more than April. How much does Ben receive?

Solution

Cate gets $6-2=4$ shares more than April.

So 4 shares = \$30

1 share = \$7.50

Ben gets 3 shares = 3 × £7.50 = \$22.50

Solving problems with scales

A model of a helicopter is 40 cm long.

The real helicopter is 10.8 m long.

Work out the scale of the model. Give your answer in the form $1:n$

Solution

Model is 40 cm = 0.4 m

Ratio of model to real helicopter is 0.4 : 10.8 → (÷0.4) → 1 : 27

Exam tip

Read the question carefully! Cate doesn't receive \$30. Always start by finding the value of 1 'share'.

Remember

Check your answer.

April gets 2 × \$7.50 = \$15

Cate gets 6 × \$7.50 = \$45

\$45 − \$15 = \$30 ✓

Watch out!

You need to make sure you convert to the same units before you work out the ratio.

Exam-style questions

1 Usain and Yohan play some games of table tennis.
Each game is won by either Usain or Yohan.
The ratio of the number of times Usain wins to the number of times Yohan wins is 7 : 3
Usain won 28 more games than Yohan.
How many games did they play? [3]

2 A rectangle has a perimeter of 64 cm.
The ratio of the rectangle's width to its length is 3 : 5
Find the area of the rectangle. [3]

3 Dinesh has a map with a scale of 1 : 25 000
His house is 7.2 km from his school.
Work out the distance on the map, in centimetres, between Dinesh's house and his school. [3]

Short answers on page 132

Full worked solutions online

CHECKED ANSWERS ONLINE

Proportion

REVISED

Key fact

Two quantities are in direct proportion when doubling one quantity means the other also doubles.

Examples: *recipe problems: amount of butter used/number of cakes made*
currency conversion problems: £1 = $1.41 so $360 = £??

You can use scaling to solve problems involving direct proportion.

Worked examples

Solving recipe problems

Here are the ingredients to make 12 flapjacks.

225 g oats 75 g sugar 50 g butter 3 tablespoons golden syrup

a How many grams of oats are needed to make 20 flapjacks?
b Aarav made some flapjacks for a party. He used 125 g of butter. How many flapjacks did Aarav make?

Solution

a
12 flapjacks need 225 g of oats
÷12 ÷12
1 flapjack needs 18.75 g of oats
×20 ×20
20 flapjacks need 375 g of oats.

b
50 g of butter makes 12 flapjacks
÷50 ÷50
1 g of butter makes 0.24 flapjacks
×125 ×125
125 g of butter makes 30 flapjacks.

Solving currency problems

Barack changes $360 to pounds. The exchange rate is £1 = $1.41
Change $360 to pounds.

Solution

$1.41 = £1
÷1.41 ÷1.41
$1 = £0.709...
×360 ×360
$360 = £255.319...

So $360 is exchanged for £255.32

Remember

It is best to find out how much is needed for one first and then multiply up.

Watch out!

Don't round until you reach the final answer otherwise you may lose accuracy marks. Store the unrounded value in your calculator.

Exam tip

Give money correct to two decimal places unless the question says otherwise.

Exam-style questions

1 Here are the ingredients to make apple crumble for 6 people.

180 g flour 120 g sugar
150 g butter 0.75 kg apples

a Work out the amount of apples needed to make apple crumble for 15 people. [2]

Heston makes apple crumble for a group of people. He uses 270 g of flour.

b Work out the number of people in the group. [2]

2 Meena gets paid the same amount for each hour she works.
One week she works 35 hours and gets paid 9800 rupees.
The next week she gets paid 2870 rupees.
How long does she work that week? Give your answer in hours and minutes. [3]

Short answers on page 132
Full worked solutions online

CHECKED ANSWERS ONLINE

Percentages

REVISED

Key facts

1 Percent means per 100

Example: $12\% = \frac{12}{100} = \frac{3}{25} = 0.12$

2 To write x as a percentage of y, find $\frac{x}{y} \times 100\%$

Percentage increase (or decrease) $= \frac{\text{difference}}{\text{original}} \times 100\%$

3 To find a percentage of an amount:

- divide by 100 to write the percentage as a decimal
- multiply

Example: 12% of 60

$0.12 \times 60 = 7.2$

4 To increase an amount by a percentage:

- add the percentage to 100%
- write this percentage as a decimal
- multiply

To decrease an amount by a percentage:

- subtract the percentage from 100%
- write this percentage as a decimal
- multiply

Remember

To write a percentage as a decimal you divide by 100

Worked examples

Percentage decrease

In a sale, all the normal prices are reduced by 15%.

The normal price of a laptop is €780

Work out the sale price of the laptop.

Solution

$100\% - 15\% = 85\%$ and $0.85 \times €780 = €663$

Finding a percentage increase

Last year, Peter's monthly pay was €1650

This year, Peter's monthly pay is €1780

Find the percentage increase in Peter's pay.

Solution

Peter's pay has increased by €1780 – €1650 = €130

$\%\text{ increase} = \frac{130}{1650} \times 100\% = 7.878...\% = 7.88\%$ (to 2 d.p.)

Exam tip

It is more efficient to use a multiplier to work out percentage change.

Watch out!

Make sure you use the right value in the denominator. Peter's original pay was €1650, so this is the value you divide by.

Exam-style questions

1 Sam takes his family out for a meal.

The meal costs $236.70 plus a service charge of 10%.

Find the total cost of the meal. [2]

2 On 1 January 2018, Switzerland had a population of 8.54 million people.

1.96 million of these people spoke French.

a What percentage, to the nearest 1%, of the Swiss population spoke French? [2]

The population of Switzerland is predicted to increase by 0.8% in 2018.

b Calculate the predicted population of Switzerland on 1 January 2019.

Give your answer correct to the nearest 10 000 people. [2]

Short answers on page 132

Full worked solutions online

CHECKED ANSWERS ONLINE

Further percentages

REVISED

Key facts

1 When money is placed in a savings account the bank pays **interest** at a given percentage rate **per annum** (per year). **Compound interest** is when interest is paid on the investment **and** on any interest already earned.
2 When an object **depreciates**, its value decreases by a certain percentage each year.
3 Sometimes you need to work backwards to find the original amount after a percentage change.
This is called a **reverse percentage**.
Instead of multiplying, you divide to reverse the percentage change.

Example:
\$1000 is invested at 10% compound interest.
After 1 year there is
$1.10 \times \$1000 = \$1100$
After 2 years there is
$1.10 \times \$1100 = \$1210$

Worked examples

Using compound interest

Isobel invested \$6000 for 3 years at 4% per annum compound interest.
Calculate the value of her investment at the end of 3 years.

Solution

$100\% + 4\% = 104\%$

$1.04 \times 1.04 \times 1.04 \times \$6000 = 1.04^3 \times \$6000 = \$6749.18$

Solving depreciation problems

Pablo bought a new motorbike two years ago.
After one year, the value of Pablo's motorbike had depreciated by 30%.
After two years, the value had depreciated by a further 40%.
Complete this sentence:
Pablo's motorbike is now worth ____% of its original value.

Solution

1st year: $100\% - 30\% = 70\%$ 2nd year: $100\% - 40\% = 60\%$

So value is 0.7×0.6 original value $= 0.42 \times$ original value

Pablo's motorbike is now worth 42% of its original value.

Solving reverse percentage problems

A shop reduces the normal price of its trainers by 15% in a sale.
The sale price of a pair of trainers is £76.50, calculate their normal price.

Solution

$100\% - 15\% = 85\%$. So 85% of normal price = sale price

So $0.85 \times$ normal price = £76.50

So the normal price = £76.50 ÷ 0.85 = £90

Exam tip

It is more efficient to use powers to work out compound interest or depreciation problems.
Here you multiply by 1.04 three times (once for each year) which is the same as multiplying by 1.04^3

Watch out!

Percentage changes are multiplied not added!

Remember

Check your answer:
$0.85 \times £90 = £76.50$ ✓

Exam-style questions

1 Hodder Savers Bank pays 3% interest per annum compound interest.
 a Freya invests £4000 for 2 years. Calculate the value of her investment after 2 years. [2]
 b Marty invests some money for 1 year. At the end of the year he has £324.45.
 How much did Marty invest? [2]
2 The population of penguins in a particular colony is decreasing at a rate of 15% per year.
This year there are 4299 penguins. Estimate how many penguins there were three years ago. [2]

Short answers on page 132

Full worked solutions online

CHECKED ANSWERS ONLINE

Powers and roots

REVISED

Key facts

1 **Index** (plural: **indices**) is another word for a power.

- $a^n = \underbrace{a \times a \times a \times \ldots \times a}_{n \text{ factors of } a}$
- $a^{-n} = \frac{1}{a^n}$
- $a^0 = 1$ and $a^1 = a$

Examples:

$4^3 = 4 \times 4 \times 4$

$7^{-1} = \frac{1}{7}$ (the reciprocal of 7)

$3.7^0 = 1$

2 Laws of indices

- $x^m \times x^n = x^{m+n}$
- $x^m \div x^n = x^{m-n}$
- $(x^m)^n = x^{mn}$

3 Roots

- $\sqrt{x} = x^{\frac{1}{2}}$
- $\sqrt[n]{x} = x^{\frac{1}{n}}$
- $\sqrt[n]{x^m} = x^{\frac{m}{n}}$

Remember

In $x^{\frac{m}{n}}$: the 'top' is the 'power' and the 'bottom' is the 'root'.

Worked examples

Using laws of indices

Write each of the following as a single power of 3

a $3^5 \times 9 \times 3$

b $\frac{9}{\sqrt{27}}$

Solution

a $3^5 \times 9 \times 3 = 3^5 \times 3^2 \times 3^1$
$= 3^{5+2+1} = 3^8$

b $\frac{9}{\sqrt{27}} = \frac{3^2}{\sqrt{3^3}} = \frac{3^2}{3^{\frac{3}{2}}}$
$= 3^{2-\frac{3}{2}} = 3^{\frac{1}{2}}$

Using laws of indices with letter symbols

Simplify:

a $\frac{3a^2b^3 \times 4ab^5}{2a^4b}$

b $(2a^3\sqrt{b})^4$

Solution

a $\frac{3a^2b^3 \times 4a^1b^5}{2a^4b}$
$= \frac{3 \times 4a^{2+1}b^{3+5}}{2a^4b}$
$= \frac{12a^3b^8}{2a^4b} = 6a^{3-4}b^{8-1}$
$= 6a^{-1}b^7 = \frac{6b^7}{a}$

b $(2a^3\sqrt{b})^4 = 2^4(a^3)^4(b^{\frac{1}{2}})^4$
$= 16a^{3\times4}b^{\frac{4}{2}}$
$= 16a^{12}b^2$

Exam tip

Evaluate the question and your answer on your calculator to check you are right.

Watch out!

Don't forget to multiply the indices when you find a 'power of a power'.

Exam-style questions

1 Write each of the following as a single power of 5.

a $5 \times 5 \times 5 \times 5$ [1]

b $\frac{5^3 \times 5^4}{5^2}$ [1]

c $(\sqrt{5})^6$ [1]

2 Simplify:

a $3x^2y \times 2x^3y^2$ [2]

b $\frac{(2xy^2)^3}{2x^2y^3}$ [2]

c $(\sqrt[3]{x})^6$ [2]

Short answers on page 133

Full worked solutions online

CHECKED ANSWERS ONLINE

Factors, multiples and primes

REVISED

Key facts

1 Any **integer** (whole number) can be expressed uniquely as a **product of prime factors**.

Examples: $60 = 2 \times 2 \times 3 \times 5$ and $70 = 2 \times 5 \times 7$

Finding a **product of prime factors** means find a **set of prime numbers** which **multiply** to give the given target number.

2 A **factor** is a number which divides exactly into another number.

Examples: Factors of 60 are 1, 2, 3, 4, 5, 6, 10, 12, 15, 20, 30 and 60

Factors of 70 are 1, 2, 5, 7, 10, 14, 35 and 70

The Highest Common Factor (HCF) of a pair of numbers is the highest number which is a factor of both numbers.

Example: The HCF of 60 and 70 is 10 since $2 \times 5 = 10$

3 A **multiple** is the result of multiplying a number by an integer.

Example: Multiples of 60 are 60, 120, 180, 240, 300, 360, 420, 480…

Multiples of 70 are 70, 140, 210, 280, 350, 420, 490, …

The Lowest Common Multiple (LCM) of a pair of numbers is the lowest number which is a multiple of both numbers.

Example: The LCM of 60 and 70 is 420 since $2 \times 5 \times 2 \times 3 \times 7 = 420$

Worked example

Using product of primes factors to find HCF and LCM

Find the Highest Common Factor (HCF) and the Lowest Common Multiple (LCM) of 150 and 165.

Solution

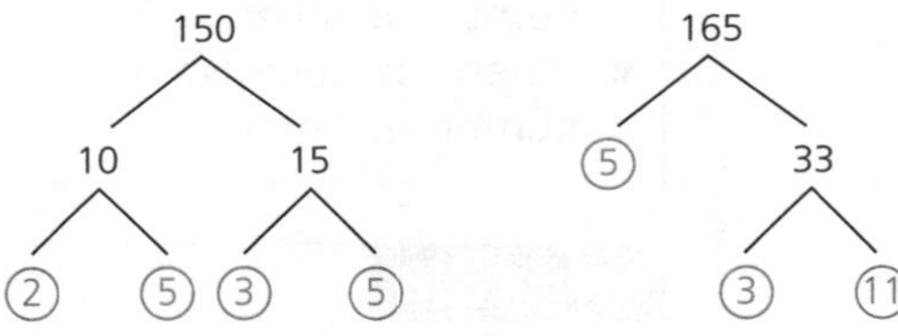

2, 5 | 3, 5 | 11

So $150 = 2 \times 3 \times 5 \times 5$ and $165 = 3 \times 5 \times 11$

So the Highest Common Factor (HCF) is $3 \times 5 = 15$

And the Lowest Common Multiple (LCM) is $3 \times 5 \times 2 \times 5 \times 11 = 1650$

Remember

You can use factor trees to help you find the product of primes.
Keep finding pairs of factors until you only have primes.

Exam tip

It is easier to use the prime factors in a **Venn diagram** to find the HCF and LCM.
For **HCF**: multiply the primes factors which are common to both numbers. This is the **intersection** of both sets.
For **LCM**: multiply the HCF by the primes factors which are **not** common to both numbers. This is the **union** of both sets.

Watch out!

The LCM is not 150×165

Exam-style questions

1 Express 1260 as a product of its prime factors. [3]

2 Given $A = 2^2 \times 3 \times 5 \times 7$ and $B = 2^5 \times 3 \times 7^2$, find:

a the HCF of A and B b the LCM of A and B. [4]

Short answers on page 133

Full worked solutions online

CHECKED ANSWERS ONLINE

Standard form

REVISED

Key fact

Standard form makes it easier to work with very large or very small numbers.

A number is in **standard form** when it is written in the form $a \times 10^n$ where n is an integer (whole number) and $1 \leq a < 10$

Examples: $3.2 \times 10^4 = 32\,000$ (multiply 3.2 by 10 four times)
$6.73 \times 10^{-3} = 0.006\,73$ (divide 6.73 by 10 three times)

Worked examples

Calculating with numbers in standard form

A space probe leaves earth and travels 4.1×10^7 km to Venus.

It then travels a further 1.2×10^8 km to Mars where it crash lands.

a How far does the probe travel altogether?
Give your answer in standard form.

b What percentage of the probe's journey is completed when the probe reaches Venus?

Solution

a Total journey is

$$4.1 \times 10^7\text{ km} + 1.2 \times 10^8\text{ km} = 41\,000\,000 + 120\,000\,000$$
$$= 161\,000\,000 = 1.61 \times 10^8\text{ km}$$

b Percentage completed at Venus is

$$\frac{\text{distance to Venus}}{\text{total distance travelled}} \times 100\% = \frac{4.1 \times 10^7}{1.61 \times 10^8} \times 100\% = 25.6\%$$

Algebra and standard form

$m = 169 \times 10^{4n}$ where n is an integer.

Express $\sqrt{m}$ in standard form.

Give your answer in terms of n, in its simplest form.

Solution

$$\sqrt{m} = \sqrt{169 \times 10^{4n}} = \sqrt{169} \times \sqrt{10^{4n}} = 13 \times \sqrt{10^{4n}}$$

Simplifying $\sqrt{10^{4n}}$ gives $(10^{4n})^{\frac{1}{2}} = 10^{4n \times \frac{1}{2}} = 10^{2n}$

So $\sqrt{m} = 13 \times 10^{2n} = 1.3 \times 10^1 \times 10^{2n} = 1.3 \times 10^{2n+1}$

Exam tip

Make sure you know how to enter numbers in standard form on your calculator. You don't need to write numbers out in full unless the question asks you to.

Remember

The power of 10 tells you how many places the decimal point is moved.

- When n is positive the number is large.
- When n is negative the number is small.

Watch out!

13×10^{2n} is not in standard form because 13 is not between 1 and 10.

Exam-style questions

1 a Write as ordinary numbers
 i 2.91×10^5
 ii 5.6×10^{-4} [2]

b Arrange these numbers in order of size. Start with the smallest number.
3.7×10^{-8} 2.9×10^{-7} 3.74×10^{-8} [2]

c Write the following in standard form.
 i 284 000
 ii 0.0284
 iii 284×10^5
 iv 28.4×10^{-5} [4]

2 The mass of an electron is 9.11×10^{-31} kg and the mass of a proton is 1.67×10^{-27} kg.
The mass of a proton is k times the mass of an electron.
Find, correct to the nearest integer, the value of k. [2]

3 $y = 4 \times 10^{2x}$ where x is an integer.
Find an expression for $y^{\frac{3}{2}}$ in terms of x.
Give your answer in standard form. [3]

Short answers on page 133

Full worked solutions online

CHECKED ANSWERS ONLINE

Degree of accuracy

REVISED

Key facts

1 The 1st significant figure (s.f.) in a number is the first non-zero digit in that number.
The 2nd significant figure is the digit immediately to the right of the 1st s.f., the next digit is the 3rd significant figure and so on.

2 You can round numbers to a given number of significant figures.
For example, to round to 2 significant figures (2 s.f.):

- Look at the **3rd significant figure**
- If it is **5 or more, round** the 2nd s.f. **up**
- If it is **4 or less, keep** the 2nd s.f. the **same**
- Replace any other digits **before** the decimal point with 0

Examples:
3865.7 = 3900 to 2 s.f.
0.001037 = 0.0010 to 2 s.f.
79.90356 = 80 to 2 s.f.

3 A rounded value lies halfway between the **lower bound, LB,** and the **upper bound, UB**.
Example: A pencil measures 7.1 cm to 1 decimal place (the nearest mm).
The actual length could be anything from 7.05 cm up to 7.15 cm.
$7.05\text{ cm} \leqslant \text{length of pencil} < 7.15\text{ cm}$
7.05 cm is the LB (least possible value) and
7.15 cm is the UB (greatest possible value) for the length of the pencil.

Worked examples

Rounding calculator answers

Find the value of $\dfrac{\sqrt{45.8}}{2.7^2 - 3.1}$ correct to 3 significant figures (s.f.).

Solution

Key

[√] [4] [5] [.] [8] [÷] [(] [2] [.] [7] [x^2] [−] [3] [.] [1] [)] [=]

into your calculator.

$\dfrac{\sqrt{45.8}}{2.7^2 - 3.1} = 1.61517772 = 1.62$ to 3 s.f.

Finding upper and lower bounds

Correct to 1 s.f., the area of a rectangle is 60 cm.
Correct to 2 s.f., the length of the rectangle is 18 cm.
Find, correct to 3 s.f., the upper and lower bounds for the width of the rectangle.

Solution

Upper and lower bounds: $55 \leqslant \text{area} < 65$ and $17.5 \leqslant \text{length} < 18.5$

UB for width $= \dfrac{65}{17.5} = 3.71$ to 3 s.f.

LB for width $= \dfrac{55}{18.5} = 2.97$ to 3 s.f.

Remember

Don't forget to use brackets.

Watch out!

- For $a - b$:
  UB is (UB a)−(LB b)
  LB is (LB a)−(UB b)
- For $\dfrac{a}{b}$:
  UB is $\dfrac{\text{UB } a}{\text{LB } b}$ LB is $\dfrac{\text{LB } a}{\text{UB } b}$

Exam-style questions

1 Use your calculator to work out the value of $\dfrac{5.8 + \sqrt{16.8}}{4.3 - 3.92}$

i Write down all the figures on your calculator display. [2]

ii Give your answer to part i correct to 3 significant figures. [1]

2 Given that $d = \dfrac{a - b}{c}$ and $a = 20$ correct to 1 s.f., $b = 12$ correct to 2 s.f. and $c = 4.37$ correct to 3 s.f., find:

i the upper bound for d, [2]

ii the lower bound for d. [2]

Short answers on page 133
Full worked solutions online

CHECKED ANSWERS ONLINE

Sets

REVISED

Key facts

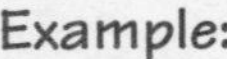

1 A set is a collection of objects.
You can use a rule to describe a set or you can list its elements (members).
Elements of a set are listed in curly brackets { }.

Example:
$A=$ {prime numbers under 10}
$A=\{2, 3, 5, 7\}$

2 $\in$ means 'is an element of'
$\notin$ means 'is not an element of'
$n\{A\}$ means the number of elements in set A
$\varnothing$ means the empty set – it is a set with no elements.
A universal set, $\mathscr{E}$, is the background set from which the members of other sets are drawn.

$3 \in A$
$8 \notin A$
A has 4 members so $n\{A\}=4$

3 Often you need to work with two or more sets.
Venn diagrams can be a useful way to represent sets.

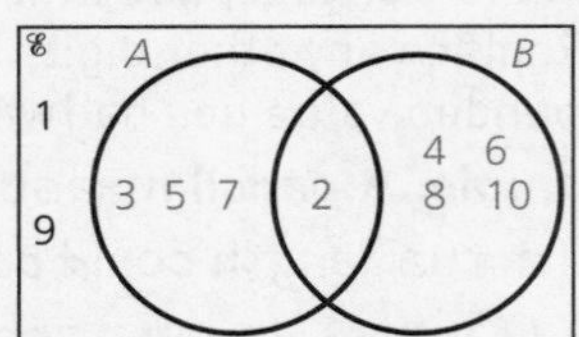

Example: $\mathscr{E}=$ {integers from 1 to 10}
$A=$ {prime numbers under 10}
$B=$ {even numbers up to 10}

- $A \cap B$ means elements that belong to both A **and** B.
This is the **intersection** of A and B. So $A \cap B=\{2\}$

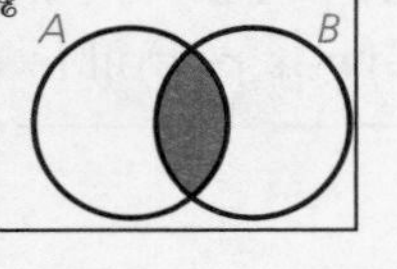

- $A \cup B$ means elements that belong to A **or** B **or** both.
This is the **union** of A and B. So $A \cup B=\{2, 3, 4, 5, 6, 7, 8, 10\}$

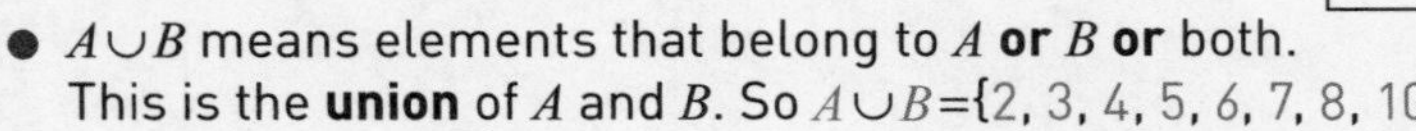

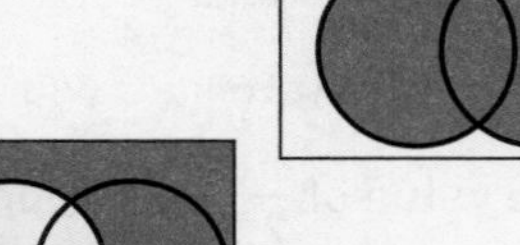

- A' is the **complement** of A.
It is all elements that are **not** in set A.
So $A'=\{1, 4, 6, 8, 9, 10\}$

- $C \subset D$ means C is a **subset** of D.
Every member of set C is also a member of set D.
On a Venn diagram, set C lies **inside** set D, so $C \cap D=C$.

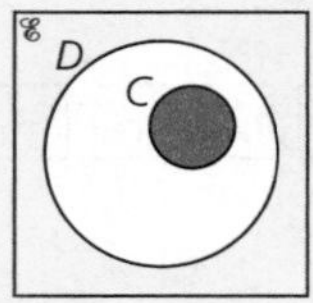

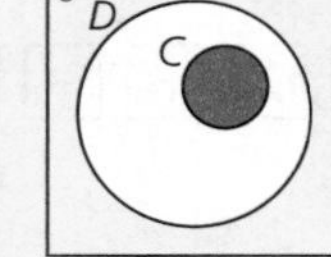

Worked examples

Understanding set notation

$A=\{a, n, g, l, e\}$, $R=\{r, h, o, m, b, u, s\}$, $S=\{s, q, u, a, r, e\}$ and $T=\{t, r, i, a, n, g, l, e\}$

a i List the members of $R \cap S$ ii Find $n(S \cup T)$

b Explain why $A \cap R=\varnothing$

c Complete these statements:
i $A \subset$ ☐ ii $h \in$ ☐

Solution

a i $R \cap S=\{r, s, u\}$

ii $S \cup T=\{s, q, u, a, r, e, t, i, n, g, l\}$ so $n(S \cup T)=11$

b $A \cap R=\varnothing$, because there are no common elements to A and R.

c i A is a subset of T, so $A \subset T$.

ii h belongs to R, so $h \in R$.

Remember

$R \cap S$ means R intersect S
List all elements common to both R and S
$S \cup T$ means S union T
List all the members of S or T or both.

Watch out!

Don't write $\varnothing$ in curly brackets!
$\{\varnothing\}$ means the set with $\varnothing$ as a member.
{ } is an alternative way to write the empty set.

Using Venn diagrams

The Venn diagram shows a universal set $\mathscr{E}$ and three sets P, Q and R.

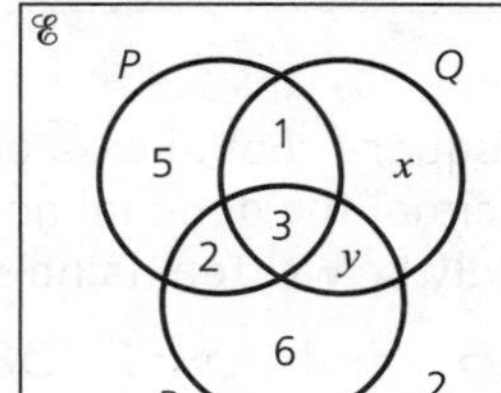

The numbers shown represent how many elements in each set.

$n(R')=12$ and $n(Q)=15$

a Find the value of x and the value of y.

b Find the value of

i $n(P\cup Q)$ ii $n(P\cup R')$.

Solution

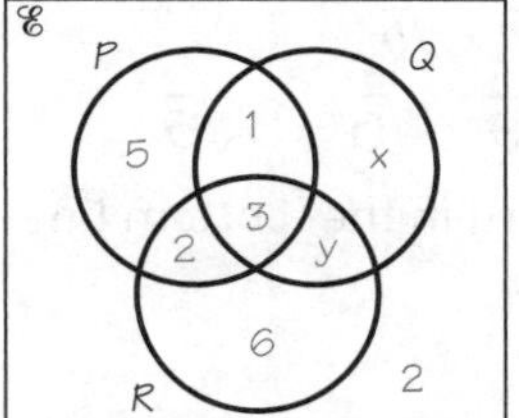

a $n(R')=12$ means there are 12 elements **not** in R

$5+1+x+2=12$

$8+x=12$, so $x=4$

$n(Q)=15$ means set Q has 15 elements

So $1+4+3+y=15$

$8+y=15$, so $y=7$

b i $n(P\cup Q)$ means how many elements in P **or** Q **or both**

$n(P\cup Q)=5+1+4+2+3+7=22$

ii $n(P\cup R')$ means how many elements in P **and not in** R

$n(P\cup R')=5+1=6$

Remember

When the circles don't intersect (so no element belongs to both sets) then $A\cap B=\varnothing$

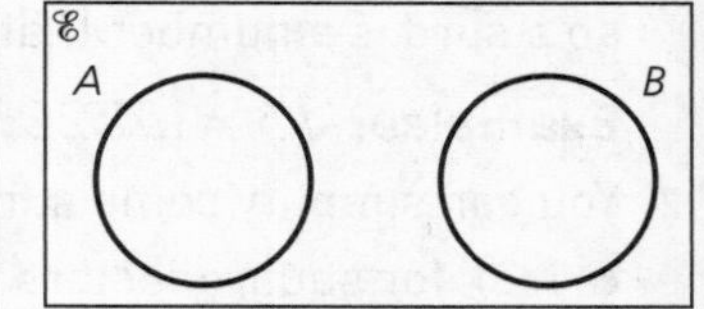

Exam-style questions

1 $\mathscr{E}$={positive integers less than 21}

A={multiples of 3}

B={factors of 18}

C={square numbers}

a List the members of the set

i $A\cap B$

ii $B\cup C$

iii $A'\cap C$ [3]

x is a member of $\mathscr{E}$.

$x\in A$ and $x\notin B$.

b Find the possible values of x [1]

$D\subset C$, $B\cap D=\varnothing$ and $n\{D\}=2$

c List the members of set D. [1]

2 The Venn diagram shows a universal set $\mathscr{E}$ and three sets A, B and C.

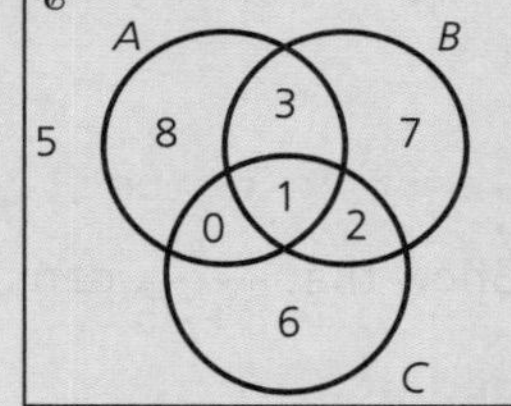

The numbers shown represent numbers of elements. Find

i $n(A\cup B)$

ii $n(B\cap C)$

iii $n(A')$

iv $n(A'\cap C)$

v $n(A'\cup B'\cup C')$ [5]

Short answers on page 133

Full worked solutions online

CHECKED ANSWERS ONLINE

Number

Surds

REVISED

Key facts

1 A surd is a number written as a square root like $\sqrt{3}$ or $\sqrt{5}$.
When a surd is a written as a decimal the decimal goes on forever and never repeats (it is irrational), so a surd is a number that can only be written using square roots.

Examples: $\sqrt{3} = 1.732050808...$ and $\sqrt{5} = 2.236067977...$

2 You can simplify some surds

- look for square factors of the number under the root
- then use the rule $\sqrt{a \times b} = \sqrt{a} \times \sqrt{b}$

Example: $\sqrt{45} = \sqrt{9 \times 5} = \sqrt{9} \times \sqrt{5} = 3\sqrt{5}$

3 To simplify a fraction with a surd in the 'bottom line' you need to rationalise the denominator.

For fractions in the form:

- $\frac{1}{\sqrt{b}}$ multiply 'top' and 'bottom' by $\sqrt{b}$
- $\frac{1}{a+\sqrt{b}}$ multiply 'top' and 'bottom' by $a-\sqrt{b}$
- $\frac{1}{a-\sqrt{b}}$ multiply 'top' and 'bottom' by $a+\sqrt{b}$

Remember

The difference of two squares (see page 27)

$$(a+\sqrt{b})(a-\sqrt{b})$$
$$= a^2 - a\sqrt{b} + a\sqrt{b} - b^2$$
$$= a^2 - b^2$$

Don't forget $\sqrt{b} \times \sqrt{b} = b$

Worked examples

Simplifying surds

Simplify $\sqrt{48} + \sqrt{75} - \sqrt{108}$

Solution

$$\sqrt{48} + \sqrt{75} - \sqrt{108} = \sqrt{16 \times 3} + \sqrt{25 \times 3} - \sqrt{36 \times 3}$$
$$= 4\sqrt{3} + 5\sqrt{3} - 6\sqrt{3} = 3\sqrt{3}$$

Rationalising the denominator

Show that $\frac{4}{3+\sqrt{5}}$ can be written as $3-\sqrt{5}$

Solution

$$\frac{4}{3+\sqrt{5}} \times \frac{3-\sqrt{5}}{3-\sqrt{5}} = \frac{4(3-\sqrt{5})}{(3+\sqrt{5})(3-\sqrt{5})} = \frac{4(3-\sqrt{5})}{3^2-(\sqrt{5})^2}$$
$$= \frac{4(3-\sqrt{5})}{9-5} = \frac{4(3-\sqrt{5})}{4} = 3-\sqrt{5}$$

Exam tip

You need to show all your working otherwise you may lose marks.

Watch out!

Don't forget to multiply the top and bottom by the same thing and then simplify. The 'bottom line' simplifies to a whole number as the middle terms cancel out.

Exam-style questions

1 Show that $(6-\sqrt{80})(3+\sqrt{5})$ can be written as $a+b\sqrt{5}$ where a and b are integers to be found. [3]

2 Show that $\frac{30}{\sqrt{12}} + \sqrt{192} = n\sqrt{3}$ where n is an integer. [3]

3 Show that $\frac{\sqrt{50}-4}{3-\sqrt{8}}$ can be written as $8+7\sqrt{2}$ [3]

Short answers on page 133

Full worked solutions online

CHECKED ANSWERS ONLINE

Review questions: Number

1 A bag contains blue and red marbles.

The marbles are either large or small.

$\frac{7}{12}$ of the marbles are blue.

$\frac{2}{3}$ of the blue marbles and $\frac{2}{5}$ of the red marbles are large.

a Show that $\frac{5}{9}$ of the marbles in the bag are large. (3 marks)

b Find the smallest possible number of marbles in the bag. (2 marks)

2 Use algebra to show that the recurring decimal $0.3\dot{0}\dot{9} = \frac{17}{55}$ (3 marks)

3 Solve $\left(\sqrt[3]{2}\right)^n = \frac{8^2}{4 \times 2^n}$ (3 marks)

Show your working clearly.

4 The highest common factor of x and y is 70

The lowest common multiple of x and y is 2100

Given $y = 420$, find the value of x.

Show your working clearly. (3 marks)

5 $y = 8 \times 10^{-6n}$

Find an expression, in terms of n, for $y^{-\frac{2}{3}}$

Give your answer in standard form. (3 marks)

6 There are 120 students in year 11 at Hodder High School.

65 students study Art.

58 students study Drama.

20 students study neither Art nor Drama.

Use a Venn diagram to work out how many students

i study both Art and Drama (2 marks)

ii study Art but not Drama (2 marks)

iii do not study Drama. (2 marks)

7 $\frac{a}{\sqrt{3}+1} = b + 3\sqrt{3}$

Find the value of a and the value of b. (4 marks)

Short answers on page 133

Full worked solutions online

CHECKED ANSWERS ONLINE

Target your revision: Algebra

Check how well you know each topic by answering these questions. If you struggle, go to the page number in brackets to revise the topic.

1 Expand and simplify expressions
Expand and simplify
- **a** $3a(2a+4)$
- **b** $2a^2b(4a-b)$
- **c** $5(3c-1)-2(4c-7)$
- **d** $(2d+3)(d+4)$
- **e** $(3e-2)^2$
- **f** $(f+1)(f-2)(f+3)$

(see page 20)

2 Factorise expressions
Factorise
- **a** a^2+4a
- **b** $6b-3b^2$
- **c** $12bc-20c^2$
- **d** $12d-6$
- **e** $36e^5f-24e^3f^2$

(see page 21)

3 Solve linear equations
Solve
- **a** $\frac{x}{4}-\frac{2x}{3}=5$
- **b** $3(2x+5)-2(2x-1)=6(x+2)$
- **c** $\frac{x+3}{2}-\frac{2x+1}{3}=\frac{x}{6}$

(see page 22)

4 Solve linear inequalities
Solve these inequalities
- **a** $4x+3 \geqslant 2x-1$
- **b** $20-4x<8$
- **c** $-5 \leqslant 2x+3<7$

Show each of your solutions on a number line.
(see page 23)

5 Find integer solutions
Given n is an integer and $-5<3n-2 \leqslant 8$ find the possible values for n.
(see page 23)

6 Solve simultaneous equations
Solve the simultaneous equations
- **a** $x+2y=4$
  $3x-4y=-18$
- **b** $7x-2y=53$
  $5x-3y=41$

(see page 24)

7 Write down formulae
There are 4 stamps in a small book of stamps.
There are 12 stamps in a large book of stamps.
Peter buys x small books of stamps and y large books of stamps.
The total number of stamps Peter buys is T.
Write down a formula for T in terms of x and y.
(see page 25)

8 Substitute values into a formula
- **a** $R=4t^2-2st$
  - **i** Work out the value of R when $s=3$ and $t=-\frac{1}{2}$
  - **ii** Work out the value of s when $R=0$ and $t=2$
- **b** $C=\frac{\sqrt{a+5}}{2b^2}$
  - **i** Work out the value of C when $a=-1$ and $b=\frac{1}{3}$
  - **ii** Work out the value of a when $C=\frac{3}{4}$ and $b=-2$

(see page 25)

9 Rearrange a formula
- **a** Make w the subject of $x=2\sqrt{w}-6$
- **b** Make x the subject of $y=\frac{x+3}{x-1}$
- **c** Given y is positive, make y the subject of $y=\sqrt{x^2+2xy^2}$

(see page 26)

10 Factorise a quadratic equation
Factorise
- **i** $16x^2-25$
- **ii** x^2-x-20
- **iii** $3x^2-16x+5$

(see pages 27–28)

11 Solve a quadratic equation by factorising
Solve the following by factorising.
- **i** $x^2-8x+16=0$
- **ii** $3x^2-7x+2=0$

(see page 29)

12 Use the quadratic formula
Solve $2x^2-3=6x$
Give your solutions correct to 3 significant figures.
Show your working clearly.
(see page 30)

13 Set up equations to solve problems
A square has sides of length $(2x-1)$ cm.
Find the value of x when
- **i** the perimeter of the square is 24 cm
- **ii** the area of the square is 25 cm^2.

(see page 31)

14 Complete the square
- **i** Write $2x^2-20x+44$ in the form $a(x+b)^2+c$
- **ii** Hence solve the equation $2x^2-20x+44=0$
  Give your answer in the form $m \pm \sqrt{n}$ where m and n are integers.
- **iii** Write down the coordinates of the turning point of the graph of $y=2x^2-20x+44$

(see page 32)

15 Solve quadratic inequalities
Solve the following inequalities.
Show each solution on a number line.
You must show clear algebraic working.
i $40 - 2x^2 \leqslant 8$
ii $x^2 + x - 2 > 0$
(see page 33)

16 Work with algebraic fractions
a Simplify fully $\frac{2x^2+7x-4}{4x^2-1}$
b Solve $\frac{2}{x+3} - \frac{1}{2x+5} = 1$
(see page 34)

17 Draw graphs to show proportion
Sketch the graph of y against x when
i y is directly proportional to x
ii y is inversely proportional to x.
(see page 35)

18 Solve problems involving proportion
P is inversely proportional to the positive square root of q. When $q = \frac{1}{4}$, $P = 10$
i Find a formula for P in terms of q.
ii Work out the value of q when $P = 25$
(see page 35)

19 Solve non-linear simultaneous equations
Solve:
$$y = 2x - 5$$
$$x^2 + y^2 = 5$$
(see page 36)

20 Use sequences and series
a The fifth term of an arithmetic series is 10
The common difference of the series is −2
Find an expression, in terms of n, for the nth term of this series.
Give your answer in its simplest form.
b The first term of a different arithmetic series is 3
The sum of the first 100 terms of this series is 25 050
Find the common difference of this series.
(see page 37)

21 Work with functions
$h(x) = 3^x$ and $f(x) = 9 - 2x$
a Find:
i $h(2)$
ii $hf\left(\frac{9}{2}\right)$
b Solve $f(x) = x$
(see page 38)

22 Find an inverse function
Given $f: x \mapsto \frac{8}{x+1}$
a State which value of x must be excluded from any domain of f.
b Find:
i $ff^{-1}(7)$
ii $f^{-1}(x)$
(see page 39)

23 Use algebra in proofs
a Prove that the square of any odd number is odd.
b Prove that the difference between the squares of any two even numbers is even.
(see page 40)

Short answers on pages 133–134
Full worked solutions online

CHECKED ANSWERS ONLINE

Expressions

REVISED

Key facts

1 In **algebra**, letters are used to represent unknown numbers called **variables** like x, y or z

A **term** is made by multiplying or dividing numbers and letters like $\frac{3xy}{2}$

An **expression** is made from one or more terms added together like $3xy - 5$

You can **simplify** an expression by combining **like terms** (terms with the **same combination** of letters).

Example: $2xy + 4xy^2 + 3xy - xy^2 = 5xy + 3xy^2$

2 You can substitute values into an expression.

Example: When $y = 3$, $5y = 5 \times 3 = 15$

3 Some expressions contain brackets.

To expand a bracket, you multiply each term by the term outside the bracket.

Example: $3x(2x - 5) = 6x^2 - 15x$

To expand a pair of brackets, multiply each term in the 2nd bracket by each term in the first.

Example: $(y+5)(y-3) = y \times y + y \times (-3) + 5 \times y + 5 \times (-3)$
$= y^2 - 3y + 5y - 15 = y^2 + 2y - 15$

4 To expand three brackets:

- first expand two brackets
- simplify the result
- multiply this result by the 3rd bracket
- then simplify.

Remember

Laws of indices (see page 10)

Worked examples

Expanding and simplifying

Expand and simplify $5(3x - 1) - 2(4x - 7)$

Solution

$5(3x - 1) - 2(4x - 7) = 15x - 5 - 8x + 14 = 7x + 9$

Expanding three brackets

Expand $(x - 2)(x + 3)^2$

Solution

Expand two brackets: $(x + 3)^2 = (x + 3)(x + 3) = x^2 + 3x + 3x + 9$
$= x^2 + 6x + 9$

Multiply by the 3rd bracket: $(x - 2)(x^2 + 6x + 9)$
$= x^3 + 6x^2 + 9x - 2x^2 - 12x - 18$

Simplifying gives: $x^3 + 4x^2 - 3x - 18$

Watch out!

Watch your signs!

Exam tip

Check your answer by substituting in a value for x. When $x = 3$:

$(x-2)(x+3)^2$
$= (3-2)(3+3)^2$
$= 1 \times 6^2 = 36$

and

$x^3 + 4x^2 - 3x - 18$
$= 3^3 + 4 \times 3^2 - 3 \times 3 - 18$
$= 27 + 36 - 9 - 18$
$= 36$ ✓

Exam-style questions

1 Expand

a $2x(3 - 5x)$ [1] b $3x^2y(4x + 6y)$ [2]

2 Expand and simplify

a $3(5x + 6) - 2(4x - 1)$ [2] b $(2x - 1)(4x + 3)$ [2]

3 Expand and simplify $(x + 4)(x + 1)(x - 2)$ [3]

Short answers on page 134

Full worked solutions online

CHECKED ANSWERS ONLINE

Factorising

REVISED

Key facts

Factorising is the reverse of expanding brackets.

To factorise an expression, you need to rewrite it using brackets.

- Look for the **highest common factor** of each term and write this outside of the bracket.
- To find the terms inside the bracket, divide each term in the original expression by the term outside the bracket.
- Check your answer by expanding the brackets.

Example:

$15x^2y + 10xy$

$5xy(\ldots + \ldots)$

$5xy(3x + 2)$

Worked examples

Factorising an expression

Factorise

i $16x^2 - 24x$

ii $9x^2y + 15xy^2 - 3xy$

Solution

i The HCF of 16 and 24 is 8 and the HCF of x^2 and x is x

So $16x^2 - 24x = 8x(2x - 3)$

Check: $8x \times 2x = 16x^2$ and $8x \times (-3) = -24x$ ✓

ii The HCF of 9, 15 and 3 is 3

and the HCF of x^2y, xy^2 and xy is xy

$9x^2y + 15xy^2 - 3xy = 3xy(3x + 5y - 1)$

Check: $3xy \times 3x = 9x^2y$ $3xy \times 5y = 15xy^2$ and $3xy \times (-1) = -3xy$ ✓

Factorising more complicated expressions

Factorise fully

i $4a^5b^2 - a^3b$

ii $6(x-3)^2 + 4(x-3)$

Solution

i The HCF of $4a^5b^2$ and a^3b is a^3b

$4a^5b^2 - a^3b = a^3b(4a^2b - 1)$

ii The HCF of $6(x-3)^2$ and $4(x-3)$ is $2(x-3)$

$$6(x-3)^2 + 4(x-3) = 2(x-3)[3(x-3)+2]$$
$$= 2(x-3)[3x-9+2]$$
$$= 2(x-3)(3x-7)$$

Watch out!

The number of terms inside the bracket should equal the number of terms in the original expression.

Watch out!

Take care with quadratic expressions such as:

- $x^2 + 5x + 6$
- $4x^2 + 1$
- $2x^2 + x - 6$

See pages 27 and 28 for how to factorise these.

Remember

The laws of indices (see page 10).

Exam tip

Sometimes you are asked to 'factorise fully' or 'factorise completely' – this just means the terms inside the bracket shouldn't have any common factors.

Exam-style questions

1 Factorise fully

a $15cd + 18de$ [1]

b $20y^2 - 5y$ [1]

2 Factorise fully $18st^2 - 36s^2t + 30st$ [2]

3 Factorise fully $8c^6d^2 + 12c^3d^4$ [2]

4 Factorise fully $8(x-5)^2 + 12(x-5)$ [2]

Short answers on page 134

Full worked solutions online

CHECKED ANSWERS ONLINE

Solving linear equations

REVISED

Key facts

1 An **equation** says two expressions are equal.
A **linear equation** contains no powers or roots of the variable (e.g. x or y).
To solve an equation, you need to rearrange it so that you have $x =$ something

2 Some tips for solving linear equations:
- gather all the x terms on one side
- keep it **balanced** – do the SAME to BOTH sides of the equation
- multiply out any brackets and simplify
- clear any fractions by multiplying EVERY term by the lowest common multiple (**LCM**) of their denominators (bottoms).

Example:
$2x+1=x+3$ is a linear equation.
$x^2+3=7$ and $9-\sqrt{x}=5$ are equations but are not linear.

Worked examples

Solving an equation with brackets

Solve $4(2x-1)-2(7-x)=6(x+1)$

Solution

$$4(2x-1)-2(7-x)=6(x+1)$$

Expand the brackets: $8x-4-14+2x=6x+6$
Simplify: $10x-18=6x+6$
Subtract $6x$ from both sides: $4x-18=6$
Add 18 to both sides: $4x=24$
Divide both sides by 4: $x=6$

Solving an equation with fractions

Solve $\dfrac{x+5}{12}+\dfrac{x-3}{4}=\dfrac{2(2x+1)}{3}$

Solution

The LCM of 3, 4 and 12 is 12, so multiply through by 12 to clear the fractions

$$\cancel{12}\times\frac{x+5}{\cancel{12}}+\cancel{12}^{3}\times\frac{x-3}{\cancel{4}}=\cancel{12}^{4}\times\frac{2(2x+1)}{\cancel{3}}$$

$$(x+5)+3(x-3)=4\times2(2x+1)$$

Expand the brackets: $x+5+3x-9=16x+8$
Simplify: $4x-4=16x+8$
Subtract $4x$ from both sides: $-4=12x+8$
Subtract 8 from both sides: $-12=12x$
Divide both sides by 12: $-1=x$, so $x=-1$

Exam tip

You can check you are right by substituting your answer back into the original equation.

Watch out!

You must multiply **every** term by the **same** number. Imagine baking a cake – you wouldn't just double some of the ingredients!

Remember

You don't have to end up with x on the left-hand side! But you might find it easier to swap the sides over:

$16x+8=4x-4$

Exam-style questions

You must **show clear algebraic working** when answering these questions.

1 Solve
- a $2(5-x)=20$ [2]
- b $\dfrac{4x+3}{5}=\dfrac{2x+1}{2}$ [3]

2 Solve
- a $4(3x-2)=3x-5$ [3]
- b $\dfrac{6}{x}=\dfrac{4}{2x-1}$ [4]

Short answers on page 134
Full worked solutions online

CHECKED ANSWERS ONLINE

Solving inequalities

REVISED

Key facts

1 An inequality says that that two values or expressions are not equal.
Examples: $x<2$ means x is less than 2
$-2 \leqslant x$ means x is greater than or equal to -2
$-3 \leqslant x < 2$ means x is greater than or equal to -3 and less than 2

2 You can **solve an inequality** in a similar way to solving an equation:
- use the inequality sign instead of =
- when you multiply or divide by a negative number, reverse the inequality.

Examples: ① $x+3<5$ ② $-2x \leqslant 6$
$x<2$ $x \geqslant -3$

3 The **solution** to an inequality is a **range of values** rather than just one value. You can represent the solution on a **number line**.
This **number line** shows $-3 \leqslant x < 2$
- ● means -3 is included
- ○ means 2 is not included

–4 –3 –2 –1 0 1 2 3 4

4 To find the integer values which satisfy an inequality, list all the whole numbers for which the inequality is true.
Example: integer values that satisfy $-3 \leqslant x < 2$ are $-3, -2, -1, 0$ and 1

Worked examples

Solving an inequality

Solve the inequality $\frac{4-2x}{3}<6$

Solution

$$\frac{4-2x}{3}<6$$

Multiply both sides by 3: $4-2x<18$

Subtract 4 from both sides: $-2x<14$

Divide both sides by -2: $x>-7$

Finding integer values

n is an integer. Find all the values of n that satisfy $-5<2n-3 \leqslant 3$

Solution

$$-5<2n-3 \leqslant 3$$

Add 3 to each side: $-2<2n \leqslant 6$

Divide each side by 2: $-1<n \leqslant 3$

n is an integer, so $n=0, 1, 2$ or 3:

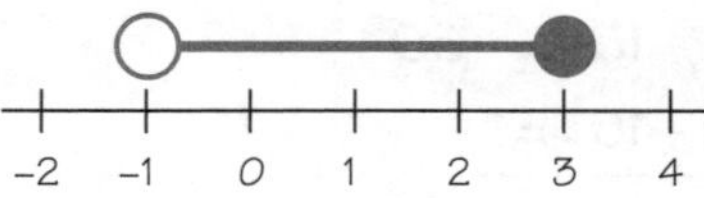

Watch out!

Keep the inequality sign pointing the same way. Only flip the inequality sign if you × or ÷ by a negative.

Exam tip

You can test a value to check your answer.
2 is greater than –7
When $x=2$:
$\frac{4-2\times 2}{3}<6$ ✓

Remember

0 is an integer.

Exam-style questions

1 An inequality is shown on the number line.
Write down this inequality.

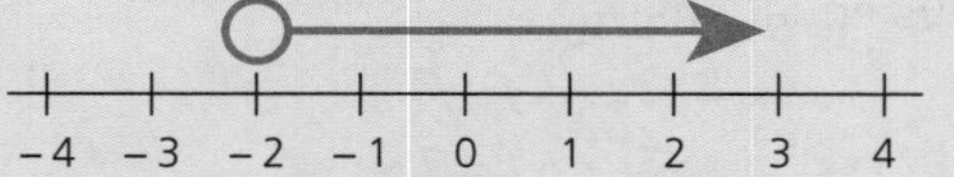

[1]

2 i Solve the inequality $-7 \leqslant 4x-1<15$ [2]
ii Write down all the integers which satisfy the inequality. [2]

3 i Solve the inequality $2(1-3x) \geqslant 26$ [2]
ii Show the solution on a number line. [2]

Short answers on page 134
Full worked solutions online

CHECKED ANSWERS ONLINE

Simultaneous equations

REVISED

Key facts

1 **Simultaneous equations** are a pair of equations both connecting two unknowns.

Example: $2x + y = 10$

$x - y = 2$

2 To solve simultaneous equations, you need to **add** or **subtract** the equations to **eliminate** one of the unknowns.

You can eliminate an unknown if both equations have the same amount of that unknown. Look at the signs of that unknown:

Different signs – add

Same signs – Subtract

3 Sometimes you need to multiply one or both equations by a suitable number to get the same amount of one unknown in both equations.

4 Once you have found the value of one unknown, substitute it into either equation to found the value of the other unknown.

Example:

$$\begin{array}{rl} & 2x + y = 10 \\ + & x - y = 2 \\ \hline & 3x \quad = 12, \text{ so } \underline{x = 4} \end{array}$$

Use $x - y = 2$ to find y:

$4 - y = 2$ so $\underline{y = 2}$

Worked examples

Solving simultaneous equations by elimination

Solve the simultaneous equations $5m + 3n = 25$

$5m - n = 5$

Solution

$5m$ has the same sign in both equations so subtract.

$$\begin{array}{rl} & 5m + 3n = 25 \\ \text{Subtract} & 5m - n = 5 \\ \hline & \quad 4n = 20 \quad \text{so } \underline{n = 5} \end{array}$$

Use $5m - n = 5$ to find the value of m:

$5m - 5 = 5$, so $5m = 10$ and $\underline{m = 2}$

Solving harder simultaneous equations

Solve the simultaneous equations $3s - 2t = -12$

$4s + 5t = 7$

Solution

To eliminate t, multiply **both** equations to get $10t$ in both.

$3s - 2t = -12$ $\times$by 5 $15s - 10t = -60$

$4s + 5t = 7$ $\times$by 2 Add $8s + 10t = 14$

$23s \quad = -46$ so $\underline{s = -2}$

$4s + 5t = 7$, so $-8 + 5t = 7$ which gives $5t = 15$ and $\underline{t = 3}$

Watch out!

It is not the signs in the 'middle' that tells you whether to add or subtract, it is the sign of the unknown you want to eliminate!

Remember

Take care when you subtract:

$3n - (-n) = 4n$

Exam tip

Don't forget to find the values of both unknowns. Check your answer works for both equations.

Exam-style questions

Showing clear algebraic working, solve the following simultaneous equations:

1 $3a - 2b = 8$ [2]

$4a + 2b = 13$

2 $2c - 4d = -22$ [3]

$3c - 2d = -17$

3 $3e - 4f = 8$ [3]

$5e + 6f = 7$

Short answers on page 134

Full worked solutions online

CHECKED ANSWERS ONLINE

Formulae

REVISED

Algebra

Key facts

1 A **formula** (plural: **formulae**) is a mathematical rule for working something out.
A formula has an equals sign and has a single variable (letter symbol) on one side.
A formula must have more than one variable:
Example: $y = 3x + 6$ is a formula but $7 = 3x + 6$ is not a formula

2 The subject of a formula is the variable which is on its own on one side of the formula.
Examples: V is the subject of the formula for the volume of a cuboid $V = lwh$
a^2 is the subject of the formula for Pythagoras' theorem $a^2 = b^2 + c^2$

3 You can work out the value of the subject of a formula by **substituting** the values of all the other variables into the formula.

Worked examples

Substituting into a formula

Given $A = 2b^2 - 4c$, work out the value(s) of:

i A when $b = -3$ and $c = 5$

ii b when $A = 51$ and $c = -\frac{1}{2}$

Solution

i Substituting $b = -3$ and $c = 5$ into $A = 2b^2 - 4c$ gives:

$A = 2 \times (-3)^2 - 4 \times 5$

$= 2 \times 9 - 20$

$= -2$

ii Substituting $A = 52$ and $c = -\frac{1}{2}$ into $2b^2 - 4c = A$ gives:

$2b^2 - 4 \times \left(-\frac{1}{2}\right) = 2b^2 + 2 = 52$

Solving to find b: $2b^2 = 50$

Dividing by 2: $b^2 = 25$ which gives $b = 5$ or $b = -5$

Writing down a formula

Sylvie repairs computers. She charges £p per hour plus a call out fee of £c. Find a formula for the total fee, F, that Sylvie charges for a repair that takes h hours.

Solution

Fee = £p multiplied by number of hours + £c, so $F = hp + c$

Remember

$2b^2$ means 2 lots of b^2
$(2b)^2$ means $2b$ all squared.

Exam tip

It is easier to swap the formula around when you are finding a value that is not the subject.

Watch out!

Don't forget the negative square root!

Exam tip

Check your formula works by making up some easy values for h, p and c.

Exam-style questions

1 The area of a small tile is 25 cm^2.
The area of a large tile is 100 cm^2.
Tom uses x small tiles and y large tiles to tile an area measuring A cm^2.
Find a formula for A in terms of x and y. [3]

2 $P = \frac{2x - y}{\sqrt{y}}$

i Work out the value of P when $x = 2.4 \times 10^3$ and $y = 400$ [2]

ii Work out the value of x when $P = 13.5$ and $y = \frac{1}{4}$ [3]

Short answers on page 134

Full worked solutions online

CHECKED ANSWERS ONLINE

Rearranging formulae

REVISED

Key fact

You can change the subject of a formula by rearranging it to get a different letter on its own on the left-hand side.

Here are some tips for rearranging formulae.

- Rewrite the formulae with the left and right-hand sides swapped over.
- You are aiming to get the new subject (target letter) on its own on the left-hand side – treat the formula as if it is an equation that you are 'solving' for the target letter.
- Keep the formula balanced by doing the **same** to **both sides.**

Worked examples

Rearranging a formula with a power

Make u the subject of the formula $v^2 = u^2 + 2as$

Solution

Swap sides: $u^2 + 2as = v^2$

Subtract $2as$ from both sides: $u^2 = v^2 - 2as$

You need u as the subject, not u^2, so square root both sides to give $u = \pm\sqrt{v^2 - 2as}$

Rearranging a formula where the new subject appears twice

Make x the subject of the formula $y = \sqrt{\frac{7x-3}{2x}}$

Solution

Swap sides: $\sqrt{\frac{7x-3}{2x}} = y$

Square both sides to remove the square root: $\frac{7x-3}{2x} = y^2$

Multiply both sides by $2x$: $7x - 3 = 2xy^2$

Add 3 to both sides: $7x = 2xy^2 + 3$

Subtract $2xy^2$ from both sides: $7x - 2xy^2 = 3$

Factorise: $x(7 - 2y^2) = 3$

Divide both sides by $7 - 2y^2$: $x = \frac{3}{7 - 2y^2}$

Exam tip

Read the question carefully: if u is positive only, you wouldn't need the negative square root.
So the answer would be $u = \sqrt{v^2 - 2as}$

Watch out!

You must remove the square root before you multiply by $2x$.

Remember

Factorising is a useful trick when the target letter appears more than once.

Exam-style questions

1 Make r the subject of the formula $A = 100\pi - 4\pi r^2$ where r is positive. [2]

2 Make b the subject of $3a + 2b = 5ab + 1$ [3]

3 Make s the subject of the formula $t = \frac{2s-5}{4s+3}$ [4]

4 Make y the subject of the formula $x = \sqrt{\frac{y}{2y+1}}$ [4]

Short answers on page 134

Full worked solutions online

CHECKED ANSWERS ONLINE

Factorising quadratic expressions (1)

REVISED

Key facts

1 A **quadratic expression** is in the form ax^2+bx+c where $a \neq 0$ and b and c are any numbers.

2 x^2+bx+c factorises to give $(x+?)(x+?)$

To find the missing numbers look for

- **two numbers** which **multiply** to give c
- and **add** to give b

Example: $x^2-7x+12$

Two numbers which multiply to give +12 and add to give −7 are −3 and −4

So $x^2-7x+12=(x-3)(x-4)$

3 You need to recognise these special forms:

- The **difference of two squares**

  $x^2-a^2=(x+a)(x-a)$

- A **perfect square**

  $x^2+2ax+a^2=(x+a)^2$ and $x^2-2ax+a^2=(x-a)^2$

Remember

How to expand brackets – see page 20

Worked examples

Factorising a quadratic expression

Factorise $x^2+3x-40$

Solution

$x^2+3x-40$

Two numbers which multiply to give −40 and add to give +3 are +8 and −5

So $x^2+3x-40=(x+8)(x-5)$

Recognising the difference of two squares

Factorise $100x^2-49$

Solution

$100x^2=(10x)^2$ is square and $49=7^2$ is also square.

So $100x^2-49=(10x+7)(10x-7)$

as it is the difference of two squares.

Recognising a perfect square

Factorise $x^2-8x+16$

Solution

$-8=2\times-4$ and $16=(-4)^2$

So $x^2-8x+16=(x-4)^2$ as it is a perfect square.

Watch out!

Take care with your signs!

When c is positive you need

- two positive or two negative numbers

When c is negative you need

- one positive and one negative number.

Exam tip

Remember factorising is the reverse of expanding so you can check your answer is right by expanding the brackets.

Remember

$(x-4)^2=(x-4)(x-4)$

Exam-style questions

Factorise the following expressions.

1 $64x^2-1$ [2]

2 x^2-x-56 [2]

3 $x^2-14x+45$ [2]

4 $x^2+12x+36$ [2]

Short answers on page 134

Full worked solutions online

CHECKED ANSWERS ONLINE

Factorising quadratic expressions (2)

REVISED

Key fact

To factorise a quadratic in the form ax^2+bx+c:

- find two numbers which multiply to give $a\times c$ and add to give b
- use these numbers to split the middle term
- factorise the first pair and the last pair of terms
- complete the factorisation.

Example: $6x^2-5x-4$

$a\times c=6\times(-4)=-24$ and $b=-5$

So $-8\times3=-24$ and $-8+3=-5$ ✓

$6x^2\underline{-5x}-4=6x^2\underline{+3x-8x}-4$

$6x^2+3x-8x-4=3x(2x+1)-4(2x+1)$

$6x^2-5x-4=(3x-4)(2x+1)$

Worked examples

Factorising a quadratic

Factorise $4x^2+11x-3$

Solution

$4x^2+11x-3$, so $a\times c=4\times(-3)=-12$ and $b=+11$

Two numbers that multiply to give -12 and add to give $+11$ are $+12$ and -1

Split the middle term: $4x^2\underline{+11x}-3=4x^2\underline{+12x-1x}-3$

Factorising pairs of terms:

$4x^2+12x-1x-3=4x(x+3)-1(x+3)$

So $4x^2+11x-3=(4x-1)(x+3)$

Factorising a quadratic with a common factor

Factorise fully $4x^2-22x+30$

Solution

Take out the common factor: $4x^2-22x+30=2[2x^2-11x+15]$

Factorise the quadratic: $a\times c=2\times15=30$ and $b=-11$

Two numbers that multiply to give $+30$ and add to give -11 are -6 and -5

Split the middle term: $2x^2\underline{-11x}+15=2x^2\underline{-6x-5x}+15$

Factorising pairs of terms:

$2x^2-6x-5x+15=2x(x-3)-5(x-3)$

So $2x^2-11x+15=(2x-5)(x-3)$

and $4x^2-22x+30=2(2x-5)(x-3)$

Remember

It doesn't matter if you split the middle term into $+12x-1x$ or $-1x+12x$, try it and see!
Notice that this method checks itself: when you factorise the two pairs of terms you end up with the same expression in both brackets – if you don't, you have gone wrong!

Remember

With practice you may be able to factorise quadratics like this straightaway by thinking about how the quadratic is formed when the brackets are expanded.

Watch out!

You were asked to factorise $4x^2-22x+30$ not $2x^2-11x+15$ so you must include the factor of 2 in your answer.

Exam-style questions

Factorise the following expressions.

1 $2x^2+13x+20$ [2]

2 $3x^2-13x-10$ [2]

3 $4x^2+20x+25$ [2]

4 $12x^2+2x-4$ [2]

Short answers on page 135

Full worked solutions online

CHECKED ANSWERS ONLINE

Solving quadratic equations

REVISED

Algebra

Key fact

A quadratic equation is in the form $ax^2+bx+c=0$ where $a\neq 0$ and b and c are any numbers.

A quadratic equation may have 2, 1 or 0 real solutions.
You can use factorising to solve some quadratic equations.

- Rearrange equation, if necessary, into the form $ax^2+bx+c=0$
- Factorise
- Set each bracket equal to 0 and solve to find x.

Watch out!

Not all quadratic equations can be factorised. You may have to use the quadratic formula (see page 30).

Worked examples

Solving a 2-term quadratic equation

Solve

i $4x^2-36=0$ ii $6x^2-3x=0$

Solution

i $4x^2-36=0$, so $4x^2=36$

Divide by 4: $x^2=9$; square rooting gives $x=3$ or $x=-3$

ii $6x^2-3x=0$; factorising gives $3x(2x-1)=0$

Either $3x=0$, so $x=0$ or $(2x-1)=0$, so $x=\frac{1}{2}$

So $x=0$ or $x=\frac{1}{2}$

Watch out!

Don't divide by x, otherwise you will lose the solution $x=0$

Solving a 3-term quadratic equation

i Solve $x^2-2x=24$

ii Hence solve $(y+3)^2-2(y+3)=24$

Solution

i Rearranging $x^2-2x=24$ into the form $ax^2+bx+c=0$ gives $x^2-2x-24=0$

Factorise: $(x+4)(x-6)=0$

Either $(x+4)=0$, so $x=-4$

or $(x-6)=0$, so $x=6$ and so $x=-4$ or $x=6$

ii $(y+3)^2-2(y+3)=24$ is based on $x^2-2x=24$

So let $x=y+3$, so $y=x-3$

When $x=-4$ then $y=-4-3=-7$

And when $x=6$ then $y=6-3=3$; so $y=-7$ or $y=3$

Remember

If an equation has an x^2 term, you must start by rearranging the equation so it is equal to 0

Exam tip

You can check you are right by substituting your answers back into the original equation.

Exam-style questions

1 Solve

i $12x^2-3=0$ [2] ii $4x^2=9x$ [2]

2 Solve by factorising

i $2x^2+5x-3=0$ [3] ii $4x^2-12x+9=0$ [3]

3 i Solve $x^2=3-2x$ [3]

ii Hence, or otherwise, solve $(y-4)^2=3-2(y-4)$ [2]

Short answers on page 135

Full worked solutions online

CHECKED ANSWERS ONLINE

The quadratic formula

REVISED

Key fact

You can use the quadratic formula $x = \frac{-b \pm \sqrt{b^2 - 4ac}}{2a}$ to solve a quadratic equation in the form $ax^2 + bx + c = 0$ where $a \neq 0$ and b and c are any numbers.

Worked example

Using the quadratic formula

Solve $3x^2 = 6 - 5x$

Give your solutions correct to 3 significant figures.

Show your working clearly.

Solution

Rearrange $3x^2 = 6 - 5x$ to give $3x^2 + 5x - 6 = 0$

So $a = 3$, $b = 5$ and $c = -6$

Substitute these values into $x = \frac{-b \pm \sqrt{b^2 - 4ac}}{2a}$

$$x = \frac{-5 \pm \sqrt{5^2 - 4 \times 3 \times (-6)}}{2 \times 3}$$

$$= \frac{-5 \pm \sqrt{25 + 72}}{6}$$

$$= \frac{-5 \pm \sqrt{97}}{6}$$

So $x = \frac{-5 + \sqrt{97}}{6} = 0.8081\ldots$ or $x = \frac{-5 - \sqrt{97}}{6} = -2.474\ldots$

So $x = -2.47$ or $x = 0.808$ correct to 3 s.f.

Remember

You must rearrange a quadratic equation into the form $ax^2 + bx + c = 0$ before you solve it.

Exam tip

When a question says 'give your answer correct to 3 significant figures' then you should use the quadratic formula.

Watch out!

You must show all your working, otherwise you may lose marks.

Exam-style questions

1 Solve $x^2 + 5x - 4 = 0$
  Give your solutions correct to 3 significant figures.
  Show your working clearly. [3]

2 Solve $2x^2 - 7 = x$
  Give your solutions correct to 3 significant figures.
  Show your working clearly. [3]

3 $p = 3r - 2r^2$
  i Find the value of p when $r = -2$ [1]
  ii Find the values of r when $p = -10$
  Give your solutions correct to 3 significant figures.
  Show your working clearly. [4]

Short answers on page 135

Full worked solutions online

CHECKED ANSWERS ONLINE

Setting up equations

REVISED

Key fact

You often need to set up the equation for yourself from a word problem or from a diagram.

You will be given information for both sides of the equation.

- Read the question carefully.
- Write both sides of the equation in terms of the same variable (letter).
- Solve the equation.

Worked examples

Setting up a linear equation

The diagram shows a rectangle.
All lengths are in centimetres.
The area of the rectangle is 119 cm².
Find the perimeter of the rectangle.

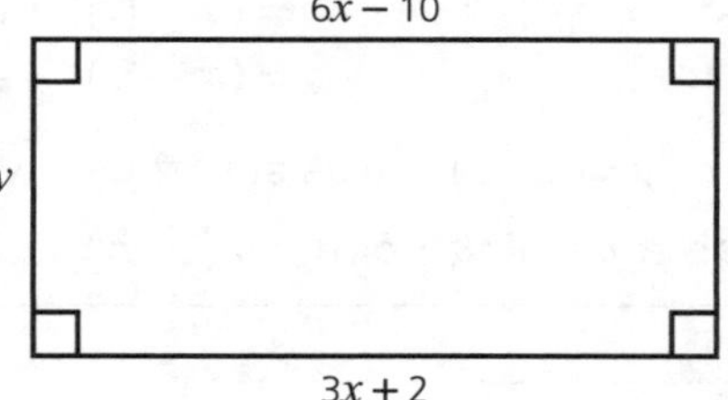

Solution

The sides marked $6x-10$ and $3x+2$ are equal in length, so:

$$6x-10=3x+2$$

Subtract $3x$ from both sides: $3x-10=2$

So $3x=12$ which gives $x=4$

So length $=3\times4+2=14$ cm (using length $=3x+2$)

The area is 119 cm², so $14\times y=119$, so $y=8.5$

So the perimeter is $2\times14+2\times8.5=45$ cm

Setting up a quadratic equation

A regular polygon has n sides and an exterior angle of $(n+2)°$

Show that $n^2+2n-360=0$ and find how many sides the polygon has.

Solution

The exterior angles of a polygon add up to 360°

360 ÷ by the number of sides = exterior angle, so $\frac{360}{n}=n+2$

Multiply both sides by n: $360=n^2+2n$

Subtract 360 from both sides: $n^2+2n-360=0$

Factorise: $(n+20)(n-18)=0$

So $n=-20$ or $n=18$; so the polygon has 18 sides.

Remember

Questions on setting up equations often involve area, perimeter and angles – see sections 4 and 5 for a reminder.

Exam tip

You were asked for the perimeter, not the value of x. Always re-read the question to make sure you have finished answering it!

Watch out!

n can't be negative as it is the number of sides so you should discard this solution.

Exam-style question

The diagram shows a square and a rectangle.
All lengths are in centimetres.

3x + 2   5x – 1   2x

a Find the width of the square when the two shapes have the same perimeter. [3]

b When the two shapes have the same area,

i show that $x^2-14x-4=0$

ii find the width of the square, giving your answer correct to 3 significant figures. [5]

Short answers on page 135

Full worked solutions online

CHECKED ANSWERS ONLINE

Completing the square

REVISED

Key facts

1 A quadratic expression is in **completed square form** when it is written in the form $(x+p)^2+q$

2 To complete the square for x^2+bx+c: Example: $x^2-10x+31$

- take the coefficient of x — -10
- halve it — -5
- square it — 25
- add and subtract this square — $\underbrace{x^2-10x+25}_{\text{this is a perfect square}}+31-25$
- factorise the **perfect square**. — $=(x-5)^2+31-25$
- Simplify — $=(x-5)^2+6$

3 The turning point of the graph of $y=(x+p)^2+q$ is at $(-p, q)$

Example: $y=(x-5)^2+6$ has a turning point at $(5, 6)$

Remember

A **perfect square** is in the form:

$x^2+2ax+a^2=(x+a)^2$

Worked examples

Using completing the square

i Write $2x^2+12x-11$ in the form $a(x+b)^2+c$

ii Hence solve the equation $2x^2+12x-11=0$
Give your answer in the form $m\pm\sqrt{n}$ where m and n are integers.

iii Write down the minimum value of y when $y=2x^2+12x-11$
and the value of x when this occurs.

Solution

i $2x^2+12x-4=2[x^2+6x-2]$

$=2[\underbrace{x^2+6x+9}_{\text{this is a perfect square}}-2-9]$

$=2[(x+3)^2-11]$

$=2(x+3)^2-22$

ii $2(x+3)^2-22=0$ so $2(x+3)^2=22$ and $(x+3)^2=11$

Square root both sides: $x+3=\pm\sqrt{11}$

Subtract 3 from both sides: $x=-3\pm\sqrt{11}$

iii $y=2(x+3)^2-22$ has a minimum value of -22 when $x=-3$

Watch out!

To complete the square on ax^2+bx+c, you need to factorise out a first.
Your answer will be in the form $a(x+p)^2+q$

Exam tip

Use the completed square form to help you answer the rest of the question.

Exam-style questions

1 i Write $x^2+8x+11$ in the form $(x+p)^2+q$ where p and q are integers. [3]

ii Hence solve the equation $x^2+8x+11=0$
Give your answer in the form $m\pm\sqrt{n}$ where m and n are integers. [2]

iii Write down the minimum value of y when $y=x^2+8x+11$ [1]

2 i Write $3x^2-24x+39$ in the form $a(x+b)^2+c$ where a, b and c are integers. [3]

ii Sketch the graph of $y=3x^2-24x+39$
Label, with coordinates, the points of intersection with the axes and the turning point. [3]

Short answers on page 135

Full worked solutions online

CHECKED ANSWERS ONLINE

Solving quadratic inequalities

REVISED

Key facts

1 The graph of a quadratic equation is a **parabola**.

The sign in front of the x^2 term tells you which way up the parabola goes.

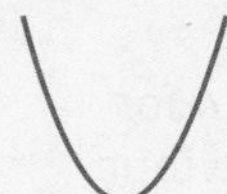

OR

Sign of x^2 is **positive**

Example: $y = x^2 + 3x - 10$

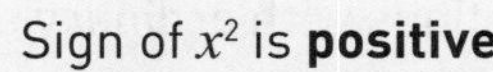

Sign of x^2 is **negative**

Example: $y = 9 - x^2$

2 To solve a **quadratic inequality** like $x^2 > 9$

- Replace the inequality sign with = and solve to find your **critical values**, a and b, where $a<b$
- The solution to the inequality is either:

  $x < a$ or $x > b$ OR $a < x < b$

  To decide which, you can either:
  - test a value between a and b to see if it satisfies the inequality
  - look at a sketch of the graph of the quadratic.

Remember

See page 47 for more about graphs of quadratic equations.

Worked examples

Solving a quadratic inequality by testing values

Solve $x^2 > 9$ and show the solution on a number line.

Solution

Solving $x^2 = 9$ gives $x = -3$ or $x = 3$, so the critical values are -3 and 3

Test a value between -3 and 3:

When $x = 2$, $2^2 < 16$ so inequality is not true when $-3 < x < 3$

So the solution is $x < -3$ or $x > 3$

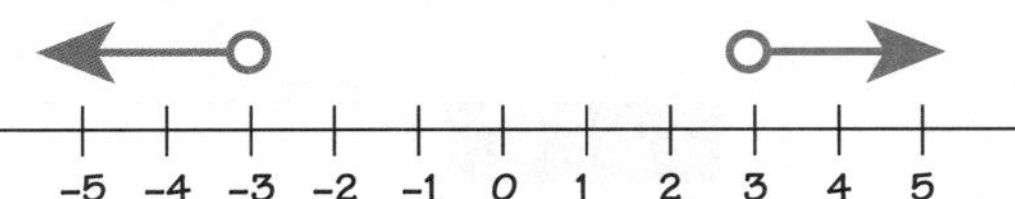

Using a graph to solve a quadratic inequality

Solve $x^2 + 3x - 10 \leqslant 0$

Solution

Solving $x^2 + 3x - 10 = 0$

Factorising gives: $(x + 5)(x - 2) = 0$

So the critical values are -5 and 2

Look at a sketch of $y = x^2 + 3x - 10$

The curve is below the x-axis between -5 and 2 so $-5 \leqslant x \leqslant 2$

Watch out!

Notice the use of the word 'or'.

x can't be both less than -3 **and** greater than 3!

Exam tip

Check: $4^2 > 9$ ✔

and $(-4)^2 > 9$ ✔

Exam tip

Two regions need two inequalities to describe them!

Remember

When the curve meets the x-axis, $x^2 + 3x - 10 = 0$

When the curve is below the x-axis, $x^2 + 3x - 10 < 0$

When the curve is above the x-axis, $x^2 + 3x - 10 > 0$

Exam-style questions

Solve the following inequalities. Show each solution on a number line. You must show clear algebraic working.

1 $5x - x^2 \leqslant 0$ [3]

2 $x^2 + 3x - 18 > 0$ [3]

3 $2x^2 - 13x + 6 < 0$ [3]

Short answers on page 135

Full worked solutions online

CHECKED ANSWERS ONLINE

Algebraic fractions

REVISED

Key facts

1 An **algebraic fraction** is a fraction which contains an **algebraic expression** in the denominator.

2 You can **simplify** some algebraic fractions by
- factorising the top and bottom of the fraction
- cancelling any common factors.

3 You can **add/subtract/multiply/divide** algebraic fractions in the same way you would ordinary numerical fractions.

Example: $\frac{2(x+5)}{x-1} \div \frac{x+5}{x+3} = \frac{2(x+5)}{x-1} \times \frac{(x+3)}{x+5} = \frac{2(x+3)}{x-1}$

Remember

See pages 3 and 4 for a reminder of the four operations with ordinary fractions.

Worked examples

Simplifying an algebraic fraction

Simplify fully $\frac{x^2-9}{x^2-2x-3}$

Solution

$$\frac{x^2-9}{x^2-2x-3} = \frac{(x+3)(x-3)}{(x-3)(x+1)} = \frac{x+3}{x+1}$$

Solving equations involving algebraic fractions

Solve $\frac{2}{x+1} - \frac{x}{x-5} = 2$

Solution

Rewrite with common denominators:

$$\frac{2(x-5)}{(x+1)(x-5)} - \frac{x(x+1)}{(x-5)(x+1)} = 2$$

Subtract: $\frac{2(x-5)-x(x+1)}{(x+1)(x-5)} = 2$

Multiply by $(x+1)(x-5)$: $2(x-5)-x(x+1) = 2(x+1)(x-5)$

Expand brackets: $2x-10-x^2-x = 2x^2-10x+2x-10$

Simplify: $x-10-x^2 = 2x^2-8x-10$

Rearrange into a quadratic: $3x^2-9x=0$

Factorise: $3x(x-3)=0$

Solve: $x=0$ or $x=3$

Watch out!

Factorise and then cancel. You can't cancel individual terms (like the x^2), you can only cancel common factors.

Remember

Make sure you remember how to factorise quadratic expressions – see pages 27 and 28 for a reminder.

Exam tip

Don't forget to check your answers:

When $x=0$

$\frac{2}{0+1} - \frac{0}{0-5} = \frac{2}{1} - 0 = 2$ ✓

When $x=3$

$\frac{2}{3+1} - \frac{3}{3-5} = \frac{2}{4} - \left(-\frac{3}{2}\right) = 2$ ✓

Exam-style questions

1 Express $\frac{2}{x+3} + \frac{1}{x-3}$ as a single fraction.
Simplify your answer fully. [3]

2 Simplify fully $\frac{x^2-1}{x^2} \div \frac{x+1}{2x}$ [3]

3 Simplify fully $\frac{2x^2-x-6}{4x^2-9}$ [4]

4 Solve $\frac{3}{2x+1} + \frac{2}{x+3} = 1$
Show clear algebraic working. [4]

Short answers on page 135

Full worked solutions online

CHECKED ANSWERS ONLINE

Proportion

REVISED

Key facts

1 $y \propto x$ means y is **directly proportional** to x.
You can write this as $y = kx$, where k is the **constant of proportionality**.
The graph of y against x is a straight line through the origin with gradient k.

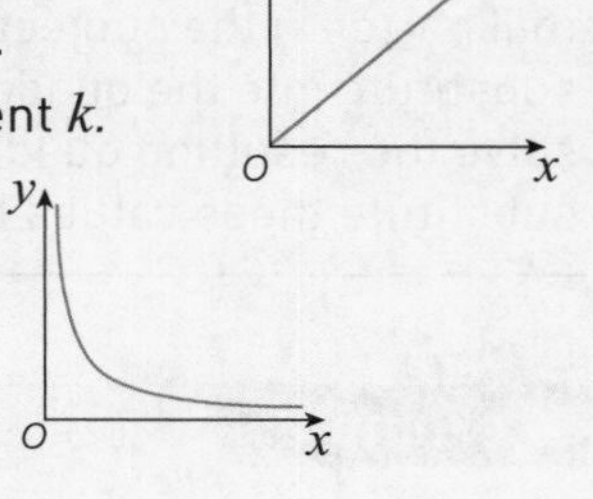

2 $y \propto \frac{1}{x}$ means y is **inversely proportional** to x.
You can write this as $y = \frac{k}{x}$
The graph of y against x is a curve.

3 Questions on proportion may involve squares, cubes or square roots.
- y is **directly proportional** to the cube of x means $y \propto x^3$ and $y = kx^3$
- y is **inversely proportional** to the square root of x means $y \propto \frac{1}{\sqrt{x}}$ and $y = \frac{k}{\sqrt{x}}$

Worked examples

Using direct proportion

P is directly proportional to the positive square root of q.
When $q = 20.25$, $P = 36$
a Find a formula for P in terms of q
b Work out the value of P when $q = 6.25$

Solution

a $P \propto \sqrt{q}$, so $P = k\sqrt{q}$
When $q = 20.25$, $P = 36$: $36 = k\sqrt{20.25} = 4.5k$
Rearranging gives $k = \frac{36}{4.5} = 8$; so $P = 8\sqrt{q}$

b When $q = 6.25$ then $P = 8\sqrt{6.25} = 20$

Using inverse proportion

The force of attraction, F, between two objects is inversely proportional to the square of the distance, d, between them.
When $d = 4$, $F = 0.05$. Find a formula for F in terms of d.

Solution

$F \propto \frac{1}{d^2}$, so $F = \frac{k}{d^2}$
When $d = 4$, $F = 0.05$: $0.05 = \frac{k}{4^2} = \frac{k}{16}$; so $k = 16 \times 0.05 = 0.8$
and hence $F = \frac{0.8}{d^2} = \frac{4}{5d^2}$

Remember

Don't forget k when you replace the proportion symbol, $\propto$, with =

Exam tip

You will be given values of x and y so that you can work out the value of k.

Watch out!

Direct proportion means when x is doubled then y is doubled.
Inverse proportion means when x is doubled then y is halved.

Exam-style questions

1 M is directly proportional to the cube of n. When $n = 2.5$, $M = 62.5$
a Find a formula for M in terms of n. [3]
b Work out the value of M when $n = 2$ [1]
c Work out the value of n, where $n > 0$, when $M = n$ [2]

2 The electrical resistance, R, of a length of wire is inversely proportional to its area, A.
a Sketch the graph of R against A. [2]
b Given that when $A = 7.5$, $R = 0.64$, find a formula for R in terms of A. [3]

Short answers on page 135

Full worked solutions online

CHECKED ANSWERS ONLINE

Non-linear simultaneous equations

REVISED

Key fact

To solve simultaneous equations where one equation is quadratic:

- make y (or x) the subject of the linear equation
- substitute into the quadratic equation
- solve the resulting quadratic to find the x (or y) value(s)
- substitute these values back into the linear equation to complete the solutions.

Worked examples

Equating the two equations

Solve the simultaneous equations $y = x^2$

$y = 6x - 9$

Solution

Both equations are equal to y, so equate them: $x^2 = 6x - 9$

Rearrange into form $ax^2 + bx + c$: $x^2 - 6x + 9 = 0$

Factorise: $(x-3)(x-3) = 0$

So $x = 3$ (the equation is a perfect square)

When $x = 3$, $y = 6 \times 3 - 9 = 9$ (using $y = 6x - 9$)

So the solution is $x = 3$ and $y = 9$

Substituting the linear equation into the quadratic

Solve the simultaneous equations $2x - y = 1$

$x^2 + y^2 = 2$

Solution

Make y the subject of $2x - y = 1$: $y = 2x - 1$ ①

Substitute for y into $x^2 + y^2 = 2$: $x^2 + (2x-1)^2 = 2$

Expand brackets: $x^2 + 4x^2 - 4x + 1 = 2$

Simplify: $5x^2 - 4x - 1 = 0$

Factorise: $(5x+1)(x-1) = 0$

So $x - 1 = 0$ which gives $x = 1$ and $y = 2 \times 1 - 1 = 1$ (from ①)

Or $5x + 1 = 0$ which gives $x = -\frac{1}{5}$ and $y = 2 \times \left(-\frac{1}{5}\right) - 1 = -\frac{7}{5}$

So the solution is $x = 1$, $y = 1$ or $x = -\frac{1}{5}$, $y = -\frac{7}{5}$

Watch out!

Do **not** substitute back into the quadratic equation to complete the pair of solutions – if you do you may find extra, invalid answers.

Exam tip

There may be one pair or two pairs of solutions. Make sure you find the x **and** y values for each pair.

Remember

See page 29 for a reminder of how to factorise and solve quadratic equations.

Exam-style questions

Solve the following simultaneous equations. Show clear algebraic working.

1 $y = x + 4$, $x^2 + y^2 = 8$ [6]

2 $y = 12x - 9$, $y = 4x^2$ [6]

3 $y - x = 3$, $x^2 + y^2 = 9$ [7]

Short answers on page 135

Full worked solutions online

CHECKED ANSWERS ONLINE

Sequences and series

REVISED

Key facts

1 A **sequence** is a list of numbers that follow a rule. — 4, 7, 10, 13, 16, ..., 151

Each number in a sequence is called a **term**.

- A **term-to-term** rule tells you how each term is related to the previous term. — First term is 4. To find the next term add 3.
- A **position-to-term** rule tells you how each term is related to its term number, n. — nth term is $3n+1$

2 In an **arithmetic sequence** the difference between each pair of terms is constant.

- The **first term** of an arithmetic sequence is a. — $a=4$
- The **common difference** is d. — $d=7-4=3$
- The **number of terms** is n. — e.g. 50th term (means $n=50$)

The ***n*th term** of an arithmetic sequence is $a+(n-1)d$ — $4+(50-1)\times 3=151$

3 An **arithmetic series** is found by adding the terms in an arithmetic sequence. — Series of first 50 terms is: $4+7+10+13+16+\ldots+151$

The sum of the first n terms, $S_n=\frac{n}{2}[2a+(n-1)d]$ — $S_{50}=\frac{50}{2}[2\times 4+(50-1)\times 3]=3875$

Worked example

Solving problems

The 4th term of an arithmetic series is 18.5

The 11th term of the same arithmetic series is 15

a Find the first term and the common difference.

b The sum of the series is 0. Work out how many terms are in the series.

Solution

a Use $a+(n-1)d$: 11th term $a+10d=15$

4th term $a+3d=18.5$ (subtract)

$7d=-3.5$ So $d=-0.5$

$a+10d=15$: when $d=-0.5$ then $a+10\times(-0.5)=15$, so $a=20$

b Use $S_n=\frac{n}{2}[2a+(n-1)d]$: $\frac{n}{2}[40+(n-1)(-0.5)]=0$

Multiply by 2 and divide by n: $40-0.5(n-1)=0$

So $0.5(n-1)=40$ which gives $n-1=80$, so $n=81$

Exam tip

Be prepared to solve simultaneous or quadratic equations when you answer questions on sequences and series.

Watch out!

Always think carefully before you divide by a letter symbol as you can't divide by 0

In this case, you know n isn't 0 as it is the number of terms.

Exam-style questions

1 The first four terms of an arithmetic sequence are 100, 93, 86, 79, ...

a Find an expression for the nth term of this sequence. [2]

b Find the sum of the first 60 terms of this sequence. [2]

2 The first term in an arithmetic series is 8 and the common difference is 3

The sum of the first n terms of the series is 2065

Find the number of terms in the series. [3]

Short answers on page 135

Full worked solutions online

CHECKED ANSWERS ONLINE

Functions

REVISED

Key facts

1 A **function** is rule that maps a number in one set to a number in another set.
Example: $f(x) = 2x + 1$ or $f : x \mapsto 2x + 1$

The **set of inputs** is called the **domain**. Domain: 1, 2, 3, 4, 5

The **set of outputs** is called the **range**. Range: 3, 5, 7, 9, 11

f(5) means substitute $x = 5$ into the function. $f(5) = 2 \times 5 + 1 = 11$

2 A **composite function, gf(x) or fg(x)** is formed when one function is applied to the output of another. Always apply the **inside function** first.

Example: $f(x) = 2x + 1$ and $g(x) = x^2$

fg(4) is $4 \xrightarrow{g} 16 \xrightarrow{f} 33$, so fg(4) = 33

gf(4) is $4 \xrightarrow{f} 9 \xrightarrow{g} 81$, so gf(4) = 81

Remember

$g(4) = 4^2 = 16$

$f(16) = 2 \times 16 + 1 = 33$

Worked examples

Using functions

The function f is defined as $f(x) = \frac{3}{2x-5}$

i Find f(1).

ii State which value of x must be excluded from any domain of f.

iii Solve the equation $f(x) = 1$

Solution

i Substitute $x = 1$ into the function: $f(1) = \frac{3}{2 \times 1 - 5} = \frac{3}{-3} = -1$

ii You can't divide by 0, $2x - 5 = 0$ gives $x = 2.5$

So $x = 2.5$ is excluded from the domain.

iii $f(x) = 1$ means $\frac{3}{2x-5} = 1$

Multiplying by $2x - 5$ gives $3 = 2x - 5$, so $2x = 8$ and $x = 4$

Finding a composite function

Given that $f(x) = 2x + 1$ and $g(x) = x^2$, find

i fg(x) ii gf(x)

Solution

i fg(x) is $x \xrightarrow{g} x^2 \xrightarrow{f} 2x^2 + 1$

ii gf(x) is $x \xrightarrow{f} 2x + 1 \xrightarrow{g} (2x + 1)^2$

Watch out!

Sometimes values are excluded from the domain. Remember you can't divide by 0 and you can't square root a negative number. Excluding values from the domain is called **restricting the domain**.

Exam tip

Think of the functions in words:
f says multiply input by 2 and add 1
g says square the input.

Remember

fg(x) is not the same as gf(x). fg(x) means apply g first, then f.
gf(x) means apply f first, then g.

Exam-style question

The functions f and g are defined as $f(x) = \sqrt{x-1}$ and $g(x) = \frac{6}{x^2+2}$

i State which values of x must be excluded from any domain of f. [1]

ii Find gf(5) [1]

iii Find gf(x)
Give your answer in its simplest form. [2]

iv Solve the equation $gf(x) = x$ [3]

Short answers on page 135
Full worked solutions online

CHECKED ANSWERS ONLINE

Inverse functions

REVISED

Key fact

The **inverse function**, f^{-1}, reverses the effect of the function f.

It maps each member of the range back to its corresponding member in the domain.

To find the inverse function:

Example: $f : x \mapsto 7 - 3x$

- replace $f(x)$ with y — $y = 7 - 3x$
- swap x and y — $x = 7 - 3y$
- rearrange to make y the subject — $y = \dfrac{7-x}{3}$
- replace y with $f^{-1}(x)$ — $f^{-1} : x \mapsto \dfrac{7-x}{3}$

Exam tip

Always check your inverse function 'undoes' the function

$2 \xrightarrow{f(2)} 1 \xrightarrow{f^{-1}(1)} 2$

Worked example

The function g is defined as $g : x \mapsto \sqrt{2x+3},\ x \geqslant 0$

i Express the inverse function g^{-1} in the form $g^{-1} : x \mapsto ...$

ii Write down the value of $gg^{-1}(15)$

iii Solve the equation $g^{-1}(x) = x$ given x is positive.

Solution

i Replace $g : x \mapsto$ with $y =$: $y = \sqrt{2x+3}$

Swap x and y: $x = \sqrt{2y+3}$

Square both sides: $x^2 = 2y + 3$

Subtract 3 from both sides: $2y = x^2 - 3$

Divide both sides by 2: $y = \dfrac{x^2-3}{2}$

Replace y with $g^{-1} : x \mapsto$: $g^{-1} : x \mapsto \dfrac{x^2-3}{2}$

ii $gg^{-1}(15) = 15$

iii $g^{-1}(x) = x$ means $\dfrac{x^2-3}{2} = x$, so $x^2 - 3 = 2x$

Subtract $2x$: $x^2 - 2x - 3 = 0$

Factorise: $(x-3)(x+1) = 0$

This gives $x = 3$ or $x = -1$, but $x \geqslant 0$ so $x = 3$

Remember

The inverse function undoes the function, so if you apply a function and then its inverse to x (or vice versa) you get back to x.

For example, if you square 15 and then square root the answer, you get back to 15.

Watch out!

Make sure you check which solutions are valid. The question states that x is positive so $x = -1$ is not a solution.

Exam-style question

The function f is defined as $f(x) = \dfrac{2x}{x-5}$

i State which value of x must be excluded from any domain of f. [1]

ii Write down the value of $ff^{-1}(3)$ [1]

iii Express the inverse function f^{-1} in the form $f^{-1}(x) = ...$ [3]

iv Solve the equation $f(x) = f^{-1}(x)$ [3]

Short answers on page 135

Full worked solutions online

CHECKED ANSWERS ONLINE

Reasoning and proof

REVISED

Key fact

When you are asked to **prove** or **show** a result is true you must show all your working.
Here are some useful starting points for questions on proof.

- When n is an **integer** (whole number): $2n$ is always even
  $2n+1$ is always odd (1 more than an even number is odd)
  $2n+1$ and $2m+1$ are two different odd numbers
- **Consecutive numbers** are numbers that follow each other in order.

Examples: 3 consecutive integers: $n, n+1, n+2$
3 consecutive even numbers: $2n, 2n+2, 2n+4$
3 consecutive odd numbers: $2n+1, 2n+3, 2n+5$

Worked examples

Proving a statement

Prove, using algebra, that the sum of any 3 consecutive even numbers is always a multiple of 6

Solution

Let the first even number $=2n$ where n is an integer

So 3 consecutive even numbers are $2n, 2n+2, 2n+4$

Finding the sum: $2n+2n+2+2n+4=6n+6$

Factorise: $=6(n+1)$

Since 6 is a factor, the sum of any 3 consecutive even numbers is a multiple of 6

Showing a result is true

Given $a>1$, show that $\frac{a}{1+\sqrt{a}}+\frac{a}{1-\sqrt{a}}=\frac{2a}{1-a}$

Solution

Rewrite with a common denominator: $\frac{a}{1+\sqrt{a}}\times\frac{1-\sqrt{a}}{1-\sqrt{a}}+\frac{a}{1-\sqrt{a}}\times\frac{1+\sqrt{a}}{1+\sqrt{a}}$

Combine the fractions: $=\frac{a(1-\sqrt{a})+a(1+\sqrt{a})}{(1-\sqrt{a})(1+\sqrt{a})}$

Expand the brackets: $=\frac{a-a\sqrt{a}+a+a\sqrt{a}}{1+\sqrt{a}-\sqrt{a}-\sqrt{a}\sqrt{a}}$

Simplify: $=\frac{2a}{1-a}$, as required.

Watch out!

The next even number is two more than the last even number.

Remember

To prove a number is a multiple of 6, you just need to show that 6 is a factor.

Exam tip

Don't forget to write a conclusion!

Exam tip

Do not start with the whole equation. Start with the more complicated side and show it simplifies to give the other side.

Exam tip

For a proof do not leave out any steps.

Exam-style questions

1 Show, using algebra, that the sum of any 4 consecutive odd numbers is always a multiple of 8 [4]

2 Given n is positive, show that $\sqrt{5n}(n\sqrt{5n}+\sqrt{20n})$ is always a multiple of 5 [3]

3 Prove, using algebra, that the sum of the first n even numbers is n^2+n [4]

Short answers on page 136
Full worked solutions online

CHECKED ANSWERS ONLINE

Review questions: Algebra

1 Prove, using algebra, that the sum of the cubes of any 3 consecutive integers is always a multiple of 3 (4 marks)

2 A stone is thrown vertically upwards from a point O.
The height above O of the stone t seconds after it was thrown from O is h metres, where $h = 16t - 5t^2$

i Find the values of t when the height of the stone is above 3 metres.
Give your answer in the form $a < t < b$ (4 marks)

ii Write $h = 16t - 5t^2$ in the form $h = p(t + q)^2 + r$
Hence find the maximum height of the stone and time taken to reach its maximum height. (5 marks)

3 P, q and r are three variables.
P is proportional to the square of q
P is also inversely proportional to the positive square root of r
$r = 9$ when $q = 5$
Find q when $r = \frac{1}{4}$
Give your answer in the form $a\sqrt{b}$ where a and b are integers. (4 marks)

4 The diagram shows a trapezium.

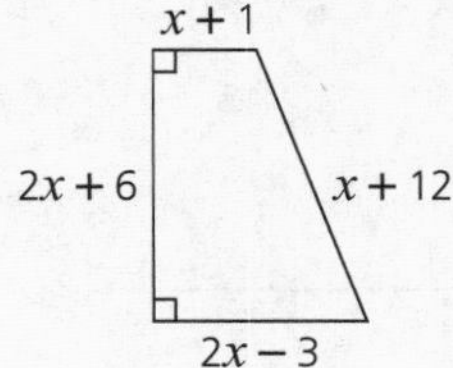

All the measurements on the diagram are in centimetres.
The area of the trapezium is 92 cm².

a Show that $3x^2 - 7x - 98 = 0$ (3 marks)

b Find the perimeter of the trapezium.
Show your working clearly. (4 marks)

5 $f(x) = x^2 - 5x - 2$ for $x > 2.5$ and $g(x) = 10 - x$

a Find $f^{-1}f(5)$ (1 mark)

b Show that $g(x) = g^{-1}(x)$ (2 marks)

c Solve $fg(x) = 0$
Give your answers correct to 3 significant figures. (4 marks)

d Solve $f(x) > g(x)$ (4 marks)

6 Peter decides to get fit.
Each week Peter runs 3 km more than he ran the week before.
In the sixth week, Peter runs 19 km.

a Find an expression for the distance Peter runs in the nth week. (2 marks)

b How far has Peter run altogether after 20 weeks? (3 marks)

Short answers on page 136

Full worked solutions online

CHECKED ANSWERS ONLINE

Target your revision: Functions and graphs

Check how well you know each topic by answering these questions. If you struggle, go to the page number in brackets to revise that topic.

1 **Find the midpoint of a line segment**

A circle has diameter AB where A is the point (–3, –1) and B is the point (3, 7)

Find the coordinates of the centre of the circle.

(see page 44)

2 **Find the gradient of a line segment**

The parallelogram $ABCD$ has vertices at the points $A(-3, -1)$, $B(0, 2)$, $C(3, 0)$ and D.

a Find the coordinates of D.

b Find the gradient of each side of $ABCD$.

(see page 44)

3 **Draw the graph of a straight line**

a On a grid, draw the graph of $x + 2y = 10$ for values of x from –2 to 12

b Find the gradient and y-intercept of your graph.

(see page 45)

4 **Use the equation of a straight line**

The equation of a line **L** is $2x - 3y = 12$

a Find the gradient of **L** and the coordinates of the y-intercept.

b Find the equation of the line with gradient 2 that passes through the point (5, –3)

(see page 45)

5 **Find equation of a parallel line**

The line **L** has equation $y = 4x + 3$

Find the equation of the straight line parallel to **L** which passes through (–2, 3)

(see page 46)

6 **Find equation of a perpendicular line**

The point A is at (–4, 2)

The point B is at (2, 5)

The point M is the midpoint of AB.

Find the equation of the line which is perpendicular to AB and passes through M.

Give your answer in the form $ax + by = c$ where a, b and c are integers.

(see page 46)

7 **Draw the graph of a quadratic equation**

a Complete the table of values for $y = 2x^2 - 4x - 3$

x	–2	–1	0	1	2	3	4
y							

b Draw the graph of $y = 2x^2 - 4x - 3$ for all values of x from $x = -2$ to $x = 4$

c Estimate the gradient of the curve $y = 2x^2 - 4x - 3$ at the point $x = 3$

d Use your graph to solve the equation $2x^2 - 4x - 3 = 5$

(see page 47)

8 **Solve equations graphically**

The diagram below shows the graph of $y = 5 + 3x - 4x^2$

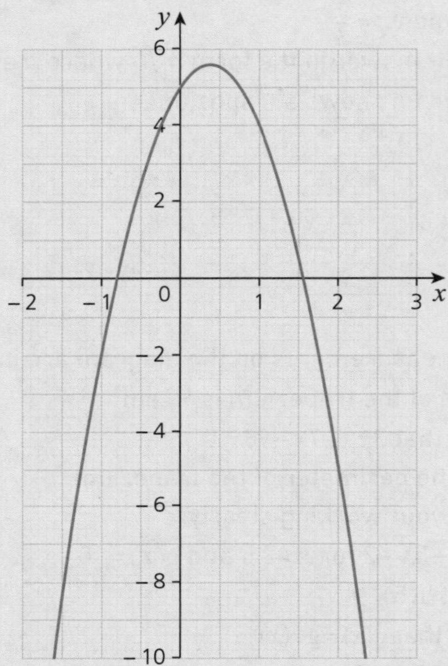

By adding suitable lines to the grid, find estimates for the solutions of:

i the simultaneous equations $y = 5 + 3x - 4x^2$ and $y = 2x - 5$

ii the equation $3x - 4x^2 = -2$

(see page 48)

9 Use inequalities to describe regions

Give the three inequalities that define the region **R**.

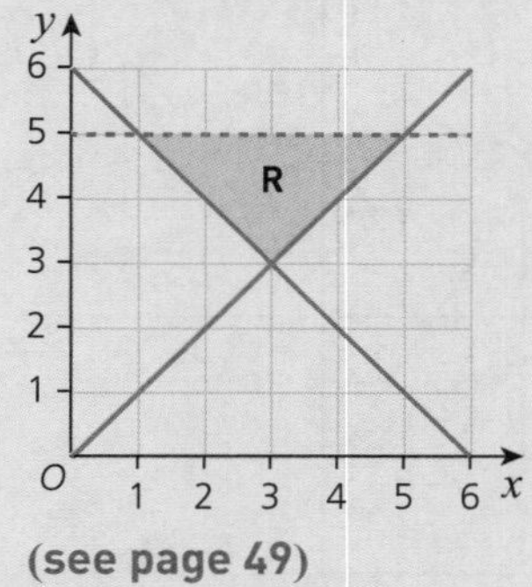

(see page 49)

10 Draw the graph of a more complicated equation

a Complete the table of values for $y = 2x^2 + \frac{1}{x} - 3,\ x \neq 0$

x	−2	−1	−0.5	−0.25	−0.1
y					

x	0.1	0.25	0.5	1	2
y					

b $f(x) = 2x^2 + \frac{1}{x} - 3$ for $-2 \leqslant x \leqslant -0.1$ and $0.1 \leqslant x \leqslant 2$

Draw the graph of $y = f(x)$

c The equation $f(x) = k$ has three solutions for $a < k < b$

Use the graph to find an estimate for the value of a and the value of b.

(see page 50)

11 Interpret real-life graphs

The graph shows the speed of an object as it moves.

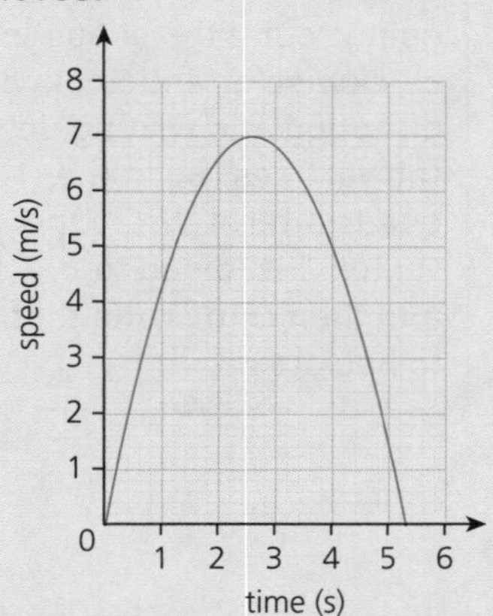

a Write down the maximum speed of the object.

b For how long does the object have a speed of 5 m/s or more?

c Estimate the acceleration of the object after 4 seconds.

(see pages 52–53)

12 Use differentiation

a Find the gradient of the curve $y = 5x^2 - \frac{3}{x^2}$ at the point where $x = \frac{1}{2}$

b Find the coordinates of the point on the curve $y = 3x^2 - 6x + 4$ where the gradient of the curve is −12

(see pages 54–55)

13 Find the coordinates of stationary points

The curve $y = 2x^2 + \frac{4}{x} - 3$ has a turning point at the point P.

Work out the coordinates of P.

(see pages 54–55)

14 Use differentiation to solve problems

A zookeeper has 288 metres of fencing.

With the 288 metres of fencing, she makes an enclosure divided into six equal, rectangular pens.

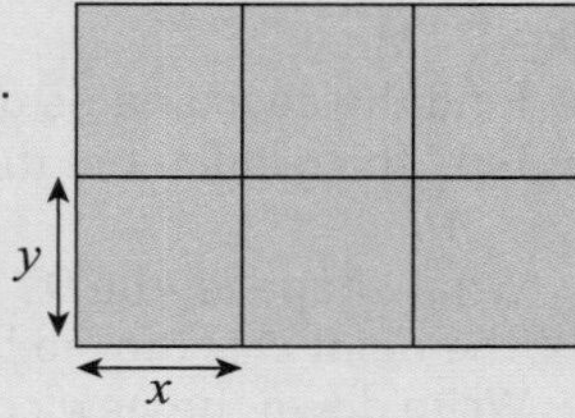

The fencing is used for the perimeter of each pen.

The length of each pen is x metres.

The width of each pen is y metres.

a Show that the total area, A m², of the enclosure is $A = 216x - \frac{27x^2}{4}$

b Find the maximum value of A.

(see page 56)

Short answers on pages 136–137
Full worked solutions online

CHECKED ANSWERS ONLINE

Coordinates

REVISED

Key facts

1 Coordinates are an ordered pair of numbers, (x, y), which give the position of a point on a grid.

2 The midpoint of two points (x_1, y_1) and (x_2, y_2) is halfway between the two points.

Midpoint $= \left(\frac{x_1 + x_2}{2}, \frac{y_1 + y_2}{2}\right)$.

3 The gradient of a line shows how steep a line is.

Gradient $= \frac{\textbf{change in } y \textbf{ coordinates}}{\textbf{change in } x \textbf{ coordinates}}$

You can think of this as **gradient** $= \frac{\textbf{rise}}{\textbf{run}}$

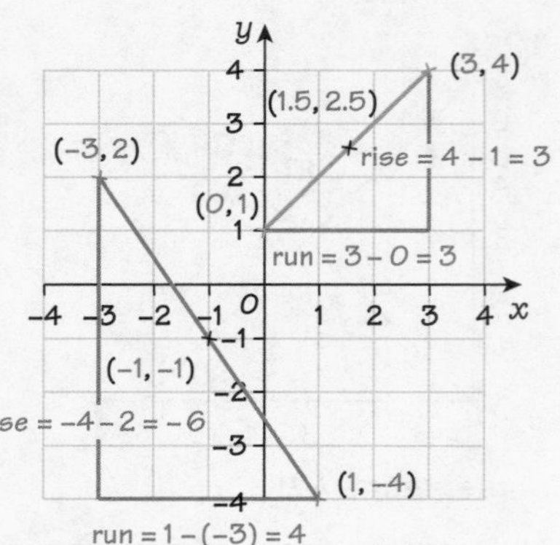

Uphill lines have a positive gradient. **Orange line**: gradient $= \frac{3}{3} = 1$

Downhill lines have a negative gradient. **Green line**: gradient $= \frac{-6}{4} = -\frac{3}{2}$

Worked examples

Solving problems using coordinates

The quadrilateral $ABCD$ has vertices at $A(-2, 1)$, $B(1, 3)$, $C(3, 1)$ and $D(-3, -3)$

a Find the coordinates of the midpoint, M, of the line segment CD.

b Find the gradient of the line segments

i AB ii CD.

c What shape is $ABCD$?

d) The point D is moved to D' so that $ABCD'$ is a parallelogram. Write down the new coordinates of D'.

Solution

a Midpoint of $C(3, 1)$ and $D(-3, -3)$ is:

$$M = \left(\frac{3 + (-3)}{2}, \frac{1 + (-3)}{2}\right) = (0, -1)$$

b i $\text{Gradient}_{AB} = \frac{\text{rise}}{\text{run}} = \frac{2}{3}$

ii $\text{Gradient}_{CD} = \frac{\text{rise}}{\text{run}} = \frac{4}{6} = \frac{2}{3}$

c $ABCD$ has one pair of parallel sides so it is a trapezium.

d A parallelogram has 2 pairs of equal, parallel sides. C is '2 right and 2 down' from B, so D' is '2 right and 2 down' from A. So D' is $(0, -1)$

Exam tip

It often helps to sketch a diagram.

Remember

- **Vertices** are the corners of a shape.
- Parallel lines have the **same** gradient.
- A **line** is infinitely long.
- A **line segment** is a part of a line.

Watch out!

A shape is named by going round the shape in a clockwise or anticlockwise direction and writing down the vertices. So in the quadrilateral $ABCD$ the vertex A is joined to B, B to C and then C to D and then D to back to A.

Exam-style question

a B is at the point (2, 3). The midpoint of AB is at (−1, 2). Work out the coordinates of A. [2]

b C is at the point $(x, -1)$. The gradient of BC is −2. Work out the value of x. [3]

c Write down the coordinates of D so that $ABCD$ is a parallelogram. [1]

Short answers on page 137

Full worked solutions online

CHECKED ANSWERS ONLINE

Straight line graphs

REVISED

Key facts

1 The equation of a straight line can be written in the form $y = mx + c$ where m is the gradient and c is the y-intercept.
Example: $y = 2x - 1$

2 To draw the graph of a straight line you can use a table of values to generate the points.

Example:

x	−1	0	1	2
$y = 2x - 1$	−3	−1	1	3

Plot the points: (−1, −3), (0, −1), (1, 1), (2, 3)

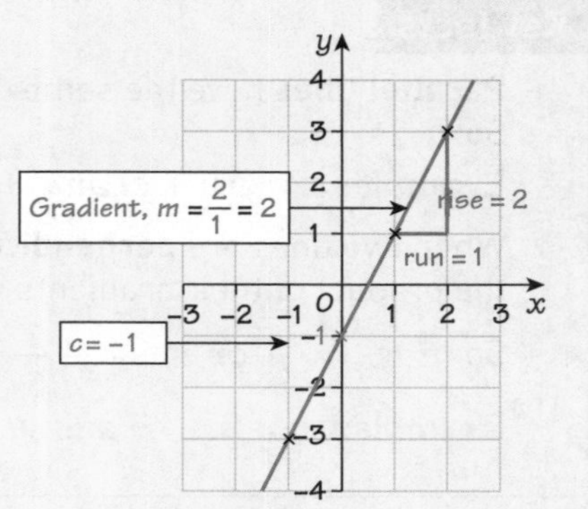

Worked examples

Drawing the graph of a straight line

On the grid, draw the graph of $2x + y = 6$ for values of x from −2 to 4

Solution

x	−2	0	2	4
$2x + y = 6$	10	6	2	−2

Using the equation of a straight line

The equation of a line **L** is $3x + 2y = 10$

Find the gradient of line **L** and write down the coordinates of the y-intercept.

Solution

Rearrange into the form $y = mx + c$: $2y = -3x + 10$,

so $y = -\frac{3}{2}x + 5$

The gradient is $-\frac{3}{2}$ and the y-intercept is (0, 5)

Finding the equation of a straight line given 2 points.

The line **L** passes through the points (1, −2) and (3, 6).

Find the equation of the line **L**.

Solution

$$\text{Gradient} = \frac{\text{change in } y \text{ coordinates}}{\text{change in } x \text{ coordinates}}$$

$$= \frac{6-(-2)}{3-1} = \frac{8}{2} = 4$$

Using $y = mx + c$ gives $y = 4x + c$

When $x = 1$, $y = -2$ so $-2 = 4 \times 1 + c$ so $c = -6$

So the equation of line L is $y = 4x - 6$

Exam tip

If you are not given a table draw up your own. You only need 2 points to draw a line but it is safer to use 3 or 4 points as a check.

$2 \times (-2) + 10 = 6$ ✓

$2 \times 4 + (-2) = 6$ ✓

Watch out!

The line must be put in the form $y = mx + c$ before you can find the gradient and y-intercept.

Remember

At the y-intercept, the x coordinate is 0

Exam tip

Use a sketch to help you.

Remember: **gradient** $= \frac{\textbf{rise}}{\textbf{run}}$

Exam-style questions

1 The line **L** passes through the points (−3, 11) and (5, −5). Find the equation of line **L**. [3]

2 The equation of a line **L** is $2x + 4y = 8$

i On a grid, draw the graph of **L** for values of x from −2 to 6 [2]

ii Find the gradient of line **L** and write down the coordinates of the y-intercept. [3]

Short answers on page 137

Full worked solutions online

CHECKED ANSWERS ONLINE

Parallel and perpendicular lines

REVISED

Key facts

1 **Parallel lines** have the same gradient.
So $m_1 = m_2$
Example: $y = 3x + 5$ and $y = 3x - 7$

2 When two lines are **perpendicular** (at right angles to each other) the product of their gradients is –1
So $m_1 m_2 = -1$ or $m_2 = -\frac{1}{m_1}$
Example: $y = 2x - 9$ and $y = -\frac{1}{2}x + 6$

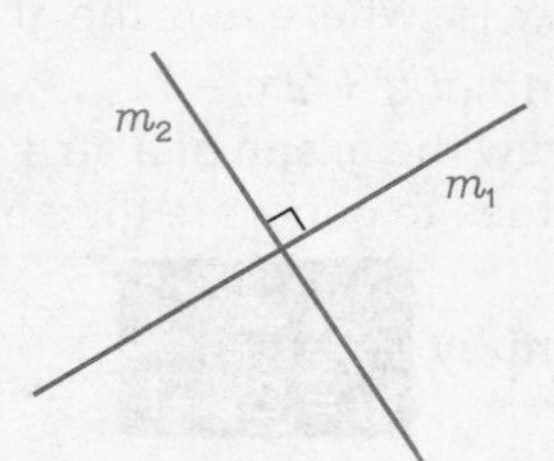

Worked examples

Find the equation of a line parallel to another

The line $\mathbf{L}_1$ passes through the points (–2, 2) and (4, –7).
The line $\mathbf{L}_2$ is parallel to $\mathbf{L}_1$ and passes through the point (2, 5).
Find the equation of $\mathbf{L}_2$. Give your answer in the form $ax + by - c = 0$

Solution

Gradient $= \frac{\text{change in } y \text{ coordinates}}{\text{change in } x \text{ coordinates}}$

$= \frac{-7-2}{4-(-2)} = \frac{-9}{6} = -\frac{3}{2}$

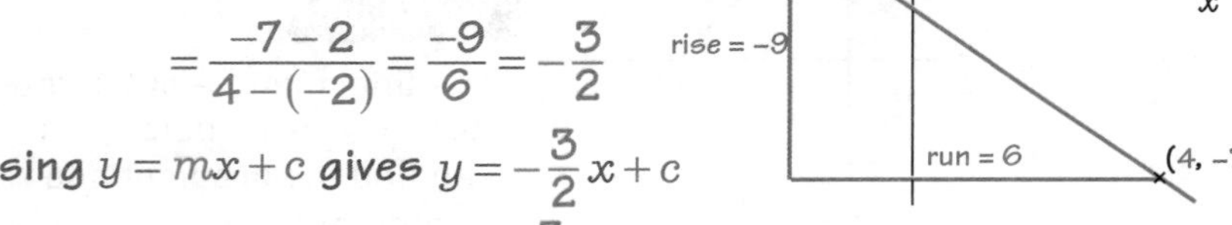

Using $y = mx + c$ gives $y = -\frac{3}{2}x + c$

When $x = 2$, $y = 5$ so $5 = -\frac{3}{2} \times 2 + c$ so $5 = -3 + c$ and $c = 8$

So the equation of $\mathbf{L}_2$ is $y = -\frac{3}{2}x + 8$

Multiply each term by 2: $2y = -3x + 16$; so $3x + 2y - 16 = 0$

Find the equation of a line perpendicular to another

The line **L** has equation $y = \frac{x}{4} - 5$. Find an equation of the line perpendicular to **L** which passes through (–1, 7).

Solution

Gradient of **L**: $y = \frac{1}{4}x - 5$ is $\frac{1}{4}$, so the perpendicular line has gradient -4

Using $y = mx + c$ gives $y = -4x + c$

When $x = -1$, $y = 7$ so $7 = -4 \times (-1) + c$

So the equation is $y = -4x + 3$

Remember

Use a sketch to help you.

Remember: **gradient** $= \frac{\text{rise}}{\text{run}}$

Exam tip

Make sure you give your answer in the right form, otherwise you may lose marks!

Watch out!

Remember the gradients of perpendicular lines multiply to give –1

$\frac{1}{4} \times (-4) = -1$ ✓

The gradients are negative reciprocals of each other.

Exam-style questions

1 The line **L** has equation $3x + 4y = 1$ The point A lies on **L** and has coordinates $(3, k)$.
 a Work out the value of k. [2]
 b Find the equation of the straight line perpendicular to **L** which passes through A. [3]

2 Line $\mathbf{L}_1$ has equation $2x - 6y = 5$ and line $\mathbf{L}_2$ has equation $y = 5 - 3x$
 a Show that $\mathbf{L}_1$ is perpendicular to $\mathbf{L}_2$. [2]
 b Write down the equation of the line parallel to $\mathbf{L}_1$ that passes through the origin. [1]

Short answers on page 137

Full worked solutions online

CHECKED ANSWERS ONLINE

Quadratic curves

REVISED

Key facts

1 A quadratic equation is in the form $y = ax^2 + bx + c$ where $a \neq 0$ and b and c are any numbers. The graph of a quadratic equation is called a **parabola**.

When a is **positive** When a is **negative**

Remember

The sign of a tells you which way up the curve is.

2 To plot the graph of a quadratic equation you use a table of values to generate the points.
Substitute each value of x into the equation of the curve to find y.

Example:

x	−3	−2	−1	0	1	2	3
$y = 2x^2 - 5$	13	3	−3	−5	−3	3	13

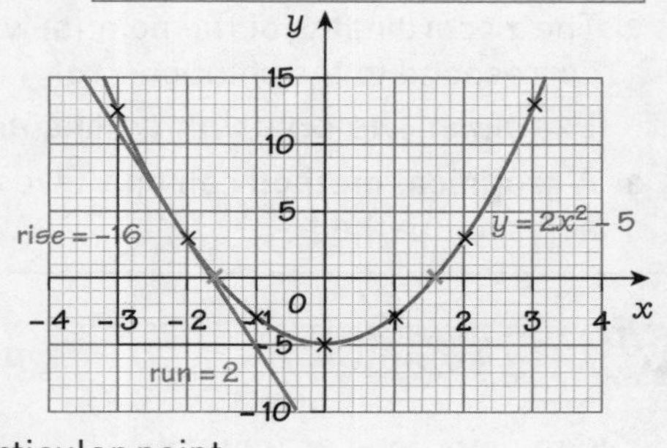

3 A tangent is a straight line which just touches the curve at a particular point. You use a tangent to estimate the gradient of the curve at that point.

Example: The purple line is the tangent at the point $x = -2$, the gradient is $\frac{-16}{2} = -8$

4 You can use the graph to **solve equations**.

Example: The orange crosses show the points where $2x^2 - 5 = 0$
So the solution to $2x^2 - 5 = 0$ is $x = -1.6$ or $x = 1.6$ to 1 d.p.

Worked example

a Draw the graph of $y = 5 + x - x^2$ for values of x between $x = -3$ and $x = 4$

b Estimate the gradient of the curve $y = 5 + x - x^2$ at the point $x = 2$

c Use your graph to solve the equation $5 + x - x^2 = -5$

Solution

a

x	−3	−2	−1	0	1	2	3	4
$y = 5 + x - x^2$	−7	−1	3	5	5	3	−1	−7

b The purple line is the tangent at the point $x = 2$, the gradient is $\frac{-9}{3} = -3$

c Add the line $y = -5$
$5 + x - x^2 = -5$ when $x = -2.7$ or $x = 3.7$ to 1 d.p.

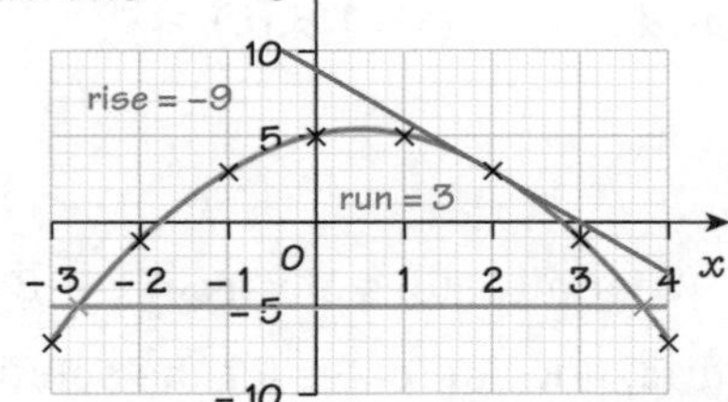

Exam tip

In the exam you will be given a table of values to complete and a grid for the graph.

Watch out!

- The graph is not flat between $x = 0$ and $x = 1$
- Join your points with a smooth curve, do not use a ruler!

Remember

Make your tangent line as long as you can – it makes it easier to work out the gradient.

Exam-style question

a Draw the graph of $y = x^2 + 2x - 4$ for values of x between $x = -5$ and $x = 3$ [3]

b Estimate the gradient of the curve $y = x^2 + 2x - 4$ at the point $x = 1$ [2]

c Use your graph to solve the equation $x^2 + 2x - 4 = 0$ [2]

Short answers on page 137

Full worked solutions online

CHECKED ANSWERS ONLINE

Graphical solution of equations

REVISED

Key facts

1. You can solve simultaneous equations graphically:
   - draw the graph of each equation
   - find the coordinates of the point(s) where the two lines **intersect** (meet).

   Example: The graphs of $y = 2x - 3$ and $y = 3 - x$ intersect at (2, 1) so the solution to the simultaneous equations is $x = 2$, $y = 1$
2. The x coordinates of the point(s) where two graphs, y_1 and y_2, intersect correspond to the solution(s) of $y_1 - y_2 = 0$

   Example: The solution to the equation $(2x - 3) - (3 - x) = 0$ is $x = 2$
3. **A graphical method** can only give an **estimate** for the solutions. It is not an exact method.

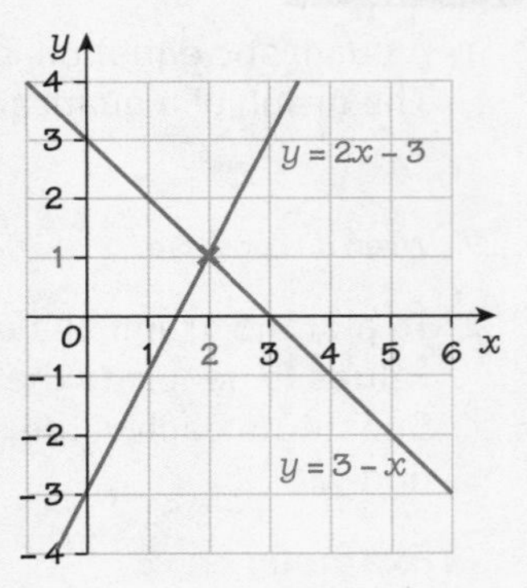

Worked example

Solving equations graphically

The diagram shows the graph of $y = x^2 + 2x - 7$

By adding suitable lines to the grid, find estimates for the solutions of

i the simultaneous equations $y = x^2 + 2x - 7$ and $y = 2x - 1$

ii the equation $x^2 + 3x - 5 = 0$

Solution

i Draw the line $y = 2x - 1$

The solutions are:

$x = -2.4$, $y = -5.9$ and $x = 2.4$, $y = 3.9$ (to 1 d.p.)

ii Start with the given equation: $x^2 + 3x - 5 = 0$

Subtract $(x + 2)$ from both sides to give $x^2 + 2x - 7 = -x - 2$

So draw the line $y = -x - 2$

The solutions are $x = -4.2$ and $x = 1.2$ (to 1 d.p.)

Remember

See pages 45 and 47 for a reminder of how to draw graphs.

Watch out!

Don't forget to find the y coordinates when you solve simultaneous equations.

Exam tip

- Use a ruler to draw a straight line.
- Read off the values as accurately as you can.

Watch out!

In (ii) you only need to give the values of x.

Exam-style question

a Draw the graph of $y = \frac{1}{2}x^2 + 3x - 2$ for all values of x from $x = -8$ to $x = 2$ [3]

b By adding suitable lines to the grid, find estimates for the solutions of

i the simultaneous equations $y = \frac{1}{2}x^2 + 3x - 2$ [3]

$y = 2x + 1$

ii the equation $\frac{1}{2}x^2 + 2x - 1 = 0$ [4]

Short answers on page 137

Full worked solutions online

CHECKED ANSWERS ONLINE

Inequalities and regions

REVISED

Key facts

You can use inequalities to show the relationship between two variables, x and y.

You can use a graph to show the region which satisfies the inequalities.

- Treat the inequality sign as though it is an =
- If necessary, rearrange the equation to make y the subject
- Draw the 'y=' line, using
  - a solid line, ——, for $y \leqslant$ or $y \geqslant$
  - a dashed line, -----, for $y <$ or $y >$
- Shade above the line for 'greater than', so $y >$ or $y \geqslant$
  Shade below the line for 'less than', so $y <$ or $y \leqslant$
- Label the shaded region **R**.

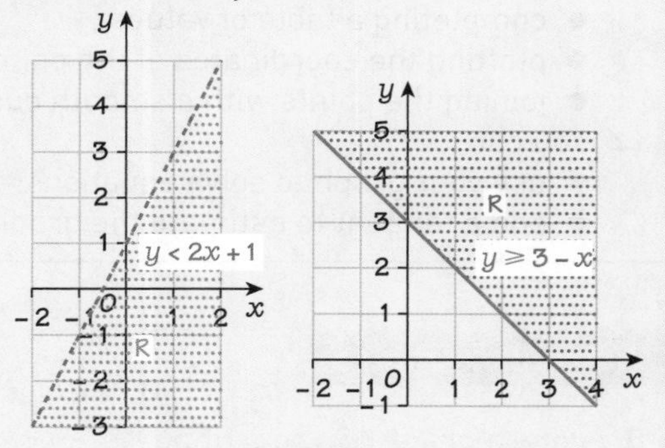

Worked examples

Showing a region on a graph

On a grid, show by shading, the region defined by the inequalities

$y \geqslant 2x - 3$ and $x + y \leqslant 6$ and $x > 1$

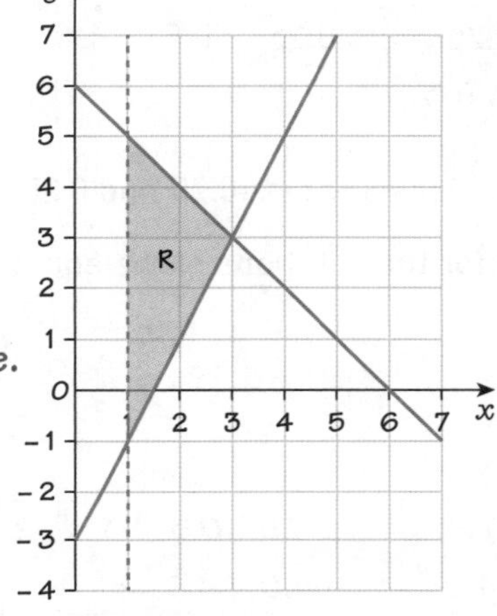

Solution

To draw the lines on a grid, find the coordinates of 2 points on each line.

$y = 2x - 3$ passes through (0, −3) and (4, 5).

$x + y = 6$ passes through (0, 6) and (6, 0).

$x = 1$ is a vertical line. Make this line dashed as it is $x > 1$

Shade the region:

- above $y = 2x - 3$ for $y \geqslant 2x - 3$
- below $x + y = 6$ for $x + y \leqslant 6$ and
- to the right of $x = 1$ for $x > 1$

Identifying a region

Give the three inequalities that define the region **R**.

Solution

The equations of the three lines are:

$y = x$, $x = 3$ and $y = -1$

The region is below $y = x$ and this line is solid so $y \leqslant x$

The region is to the left of $x = 3$ and this line is dashed so $x < 3$

The region is above $y = -1$ and this line is solid so $y \geqslant -1$

Remember

Don't forget the equations of these lines:

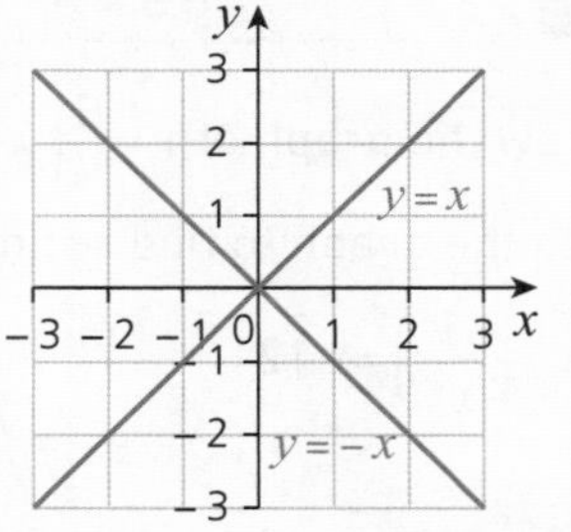

Vertical lines have equations in the form $x = a$

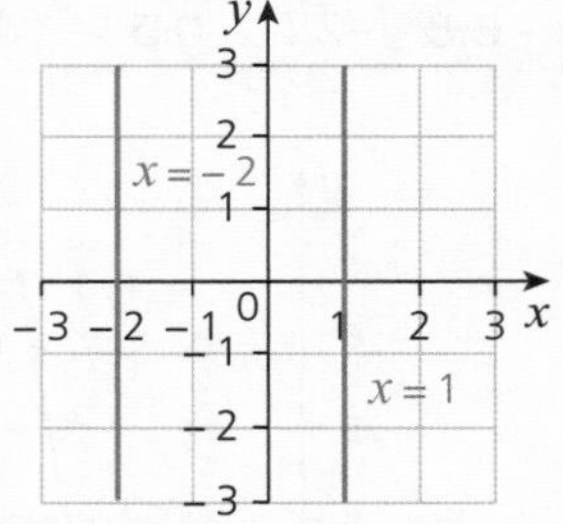

Horizontal lines have equations in the form $y = b$

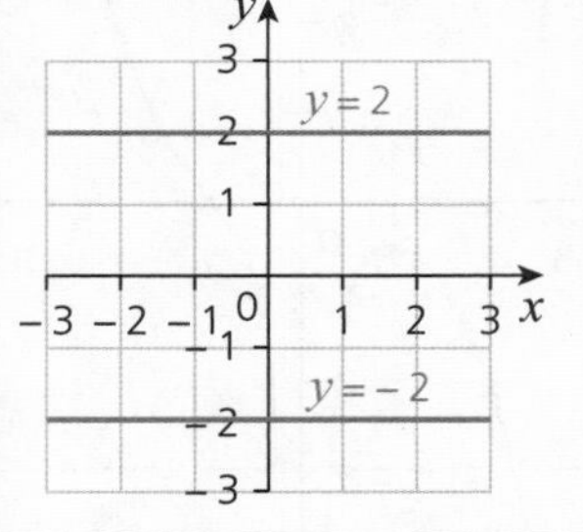

Exam-style question

i On a grid, show by shading, the region **R** defined by all three of the inequalities $y \leqslant x + 1$, $2x + y < 8$ and $y > 1$ [4]

ii The point P lies in the region **R**. Give all the possible integer coordinates of P. [2]

Short answers on page 137

Full worked solutions online

CHECKED ANSWERS ONLINE

Further graphs

REVISED

Key facts

1 You can draw the graph of a more complicated equation by:
- completing a table of values
- plotting the coordinates of the points generated from your table
- joining the points with a smooth curve.

2 You can:
- use your graph to solve equations; add a suitable line to help you
- add a tangent to estimate the gradient of the curve at a particular point.

Worked examples

Drawing a graph of a function

a Complete the table of values for $y=\frac{1}{4}\left(x^3-\frac{1}{x^2}\right)$, $x\neq 0$

x	−3	−2	−1	−0.5	−0.25		0.25	0.5	1	2	3
y	−6.8		−0.5		−4.0		−4.0	−1.0			6.7

b Draw the graph of $y=\frac{1}{4}\left(x^3-\frac{1}{x^2}\right)$ for $-3\leqslant x\leqslant -0.25$ and $0.25\leqslant x\leqslant 3$

c Use the graph to find estimates for the solutions of the equation

$$\left(x^3-\frac{1}{x^2}\right)=-12$$

Solution

a

x	−3	−2	−1	−0.5	−0.25		0.25	0.5	1	2	3
y	−6.8	−2.1	−0.5	−1.0	−4.0		−4.0	−1.0	0	1.9	6.7

b

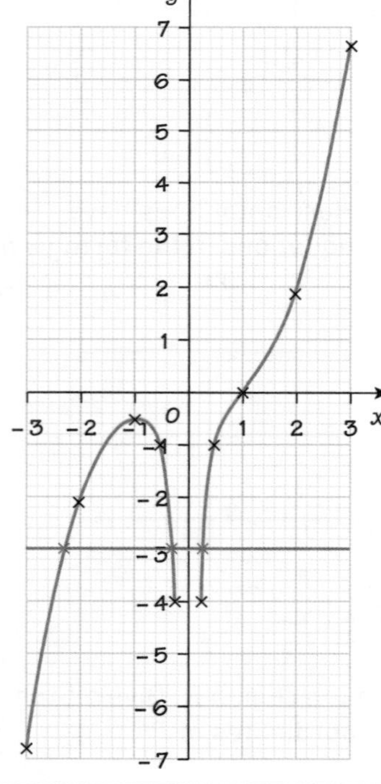

c $\left(x^3-\frac{1}{x^2}\right)=-12$

Divide both side by 4:

$$\frac{1}{4}\left(x^3-\frac{1}{x^2}\right)=-3$$

Add the line $y=-3$ to the grid.

Read off the x-values where the line crosses the curve, so the solutions are $x=-2.2$, $x=-0.3$ and $x=0.3$

Exam tip

Check you are using the equation correctly by testing a value already given to you. Use the same degree of accuracy for your answers – in this question the y-values are given correct to 1 d.p.

Watch out!

When an equation has a $\frac{1}{x}$ or a $\frac{1}{x^2}$ term, it is not defined for $x=0$ and so the curve will have two 'branches', one on either side of the y-axis. Do not join the two 'branches'! In this question, values of x between −0.25 and 0.25 have been omitted.

Remember

- **Plot** the points as **accurately** as you can.
- Join them with a **smooth curve**.
- Do **not** use a ruler to join the points.

Watch out!

Make sure you give all the solutions!

Using the graph of a function

Below is the graph of $y = f(x)$.

a Use the graph to find:

 i an estimate for the gradient of the curve $y = f(x)$ at $x = -4$

 ii the value of ff(4)

b The equation $f(x) = k$ has 3 different solutions for $a < k < b$

 Use the graph to find an estimate for the value of a and the value of b.

Solution

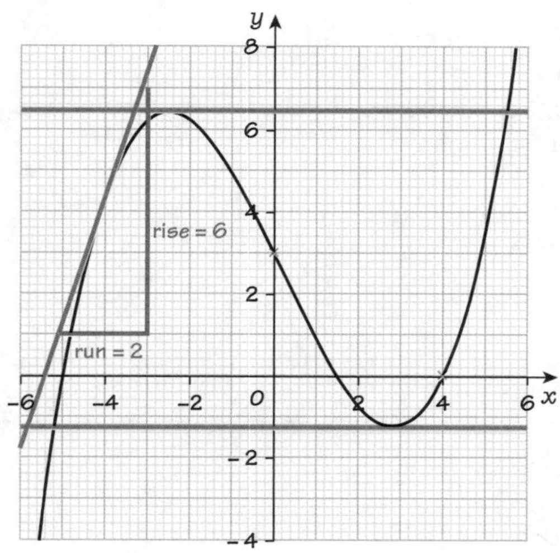

a i Draw a tangent at the point $x = -4$

 Gradient $= \dfrac{\text{rise}}{\text{run}} = \dfrac{6}{2} = 3$

 ii When $x = 4$, $f(4) = 0$ and when $x = 0$, $f(0) = 3$

 So $ff(4) = 3$

b There are 3 solutions to $f(x) = k$ for any line $y = k$ between the two green lines at $y = -1.2$ and $y = 6.5$

 So $a = -1.2$ and $b = 6.5$

Watch out!

Check the scales carefully – there may be a different scale on each axis.

Exam tip

Make sure you use a ruler when you add lines to the graph.

Remember

ff(4) means find the value of f when $x = 4$ and then find f of this value.

$4 \xrightarrow{f(4)} 0 \xrightarrow{f(0)} 3$

Exam tip

Draw the green lines to show where there are only 2 solutions.

Exam-style questions

1 Here is the graph of $y = f(x)$, $-6 \leqslant x \leqslant 6$

 a Use the graph to find an estimate for

  i the value of ff(5) [2]

  ii the solutions to the equation $f(x) = x$ [3]

 b The equation $f(x) = k$ has only two solutions for $a < k < b$. Use the graph to find an estimate for the value of a and the value of b. [2]

2 a Complete the table of values for $y = x^2 + \dfrac{2}{x} - 5$, $x \neq 0$ [2]

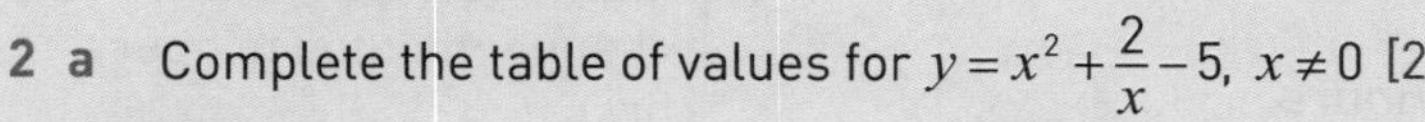

x	−3	−2	−1	−0.5	−0.3		0.3	0.5	1	2	3
y	3.3		−6		−11.6		1.8			0	

 b Draw the graph of $y = x^2 + \dfrac{2}{x} - 5$ for $-3 \leqslant x \leqslant -0.3$ and $0.3 \leqslant x \leqslant 3$ [4]

 c The gradient of the curve at $x = k$ is 0. Use the graph to find the value of k. [2]

 d Estimate the gradient of the curve at the point where $x = -1$ [3]

Short answers on pages 137–138

Full worked solutions online

CHECKED ANSWERS ONLINE

Real-life graphs

REVISED

Key facts

1 Graphs can be used to represent real-life situations.

2 A distance-time graph is used to show the distance travelled by an object over time.

- A horizontal line segment (AB) shows the object is at rest.
- Gradient = speed
- The **steeper** the line segment, the **greater the speed**.
  So the object is travelling faster in section OA than BC.
- A curve (where the gradient (i.e. speed) is changing) shows the object is **accelerating** (speeding up) or **decelerating** (slowing down).
  To estimate the acceleration at a given time, draw a tangent to the curve.

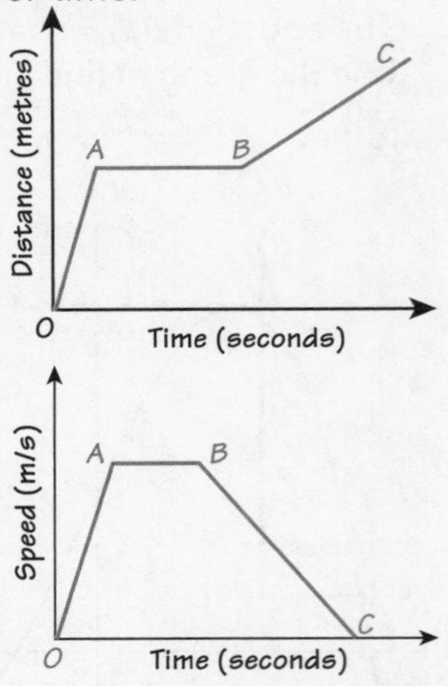

3 A **speed-time** graph is used to show how the speed of an object changes over time.

- Gradient = acceleration
- A diagonal line segment represents **constant acceleration.**
- A horizontal line segment represents **constant speed.**

4 Speed is usually measured in km/h or m/s

$$\text{speed} = \frac{\text{distance}}{\text{time}}$$

Worked examples

Using a distance-time graph

Alex leaves home at 10.00 and walks to town. The graph shows his journey.

a On the way to town, Alex stops to talk to a friend.
How long does Alex stop for?

b Find Alex's fastest walking speed.
Give your answer in km/h.

c Alex's sister, Lily, leaves home at 10.30 and jogs to town along the same route as Alex at a constant speed of 8 km/h.

i At what time does Lily overtake Alex?

ii How far from home are they?

Distance from Alex's house (km) — Time of day (10.00, 10.30, 11.00, 11.30)

Solution

a 10 small squares = 30 mins, so 1 small square = 3 mins

The horizontal segment is 3 small squares long.

So Alex stops for 9 minutes.

b Alex is walking fastest in the first part of his journey as this line is steepest.

Distance = 2 km

and time = $8 \times 3 = 24$ minutes = 0.4 hours

$$\text{speed} = \frac{\text{distance}}{\text{time}} = \frac{2}{0.4} = 5\,\text{km/h}$$

Exam tip

Check the scales carefully!

Remember

Gradient = speed
A horizontal section has a gradient of 0, and so the object is at rest.

Watch out!

Take care with your units. You are asked for km/h not km/minute so you need to change minutes to hours first.
Divide by 60 to change minutes to hours.

c Add a line for Lily's journey to the graph.

Find the point where the two graphs cross.

i Lily overtakes Alex at 10.51

ii at 2.8 km from home.

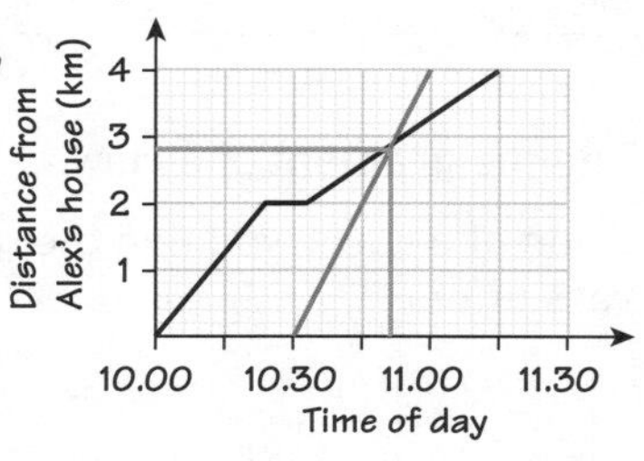

Watch out!

Lily leaves at 10.30, so her line doesn't start at the origin.
Her speed is 8 km/h so it takes her 30 minutes to travel 4 km.

Using a speed-time graph

The graph below shows the speed of an object as it moves.

a Write down the maximum speed of the object.

b For how long does the object have a speed of 6 m/s or more?

c Estimate the acceleration of the object after 1 second.

Solution

a Maximum speed = 8.8 m/s

b Speed is 6 m/s at 1.2 s and 4.2 s

4.2 − 1.2 = 3

So speed is 6 m/s or more for 3 seconds.

c acceleration = gradient of curve

Draw a tangent to the curve at t = 1

$$\text{gradient} = \frac{\text{rise}}{\text{run}} = \frac{8.2}{2} = 4.1$$

So the acceleration is $4.1\ \text{m/s}^2$.

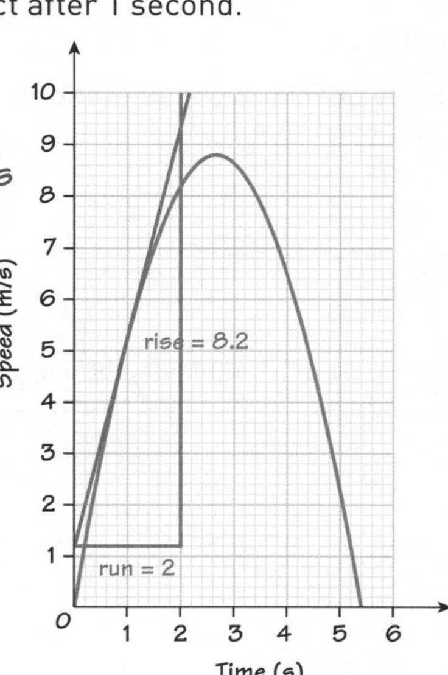

Exam tip

Always use a ruler when you add lines to a diagram.

Remember

See page 80 for more on speed.

Exam-style question

Priya walks along a path from her home to the swimming pool.

She goes for a swim and then runs back home along the same path.

Here is the distance-time graph for Priya's complete journey.

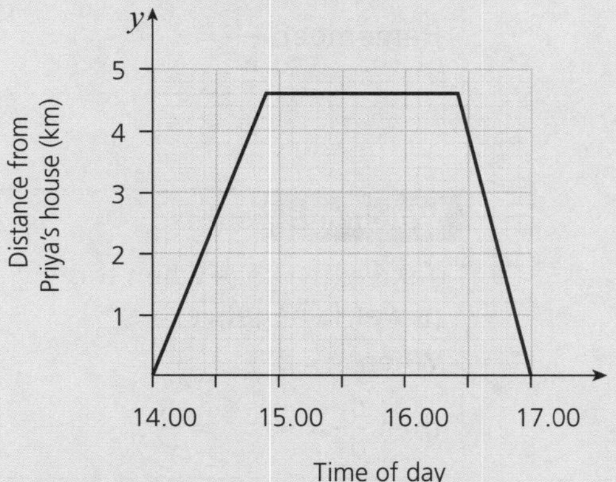

a How long does Priya stay at the pool? [1]

b What is Priya's average speed in km/h on her journey home?
Give your answer correct to 1 d.p. [2]

c Priya's brother, Sam, leaves the swimming pool at 14.30. He cycles back to Priya's home at a speed of 9.2 km/h.
How far does Sam cycle before he passes Priya on her way to the pool? [3]

Short answers on page 138
Full worked solutions online

CHECKED ANSWERS ONLINE

Differentiation

REVISED

Key facts

1 You can use **differentiation** to find the **gradient** of a curve.

When you differentiate the equation of a curve you find the **gradient function**, $\frac{dy}{dx}$

The rule for differentiating powers of x is:

$y = ax^n$ gives $\frac{dy}{dx} = anx^{n-1}$

When you differentiate a constant (a number), $\frac{dy}{dx} = 0$

Examples:

$y = x^4$, $\frac{dy}{dx} = 4x^3$

$y = 7x$, $\frac{dy}{dx} = 7$

$y = 6$, $\frac{dy}{dx} = 0$

2 To find the **gradient of a curve at a point**:

- differentiate the equation of the curve
- substitute the x coordinate into the gradient function.

3 At a **stationary point** the gradient is 0

Maximum and minimum points are special types of stationary points. Because the curve 'turns' at these points they are also called turning points.

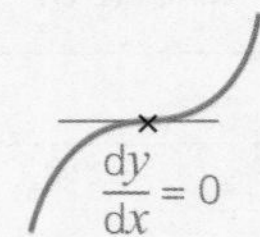

Stationary point

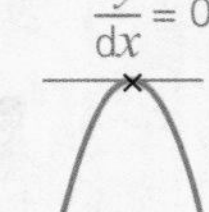

Maximum turning point

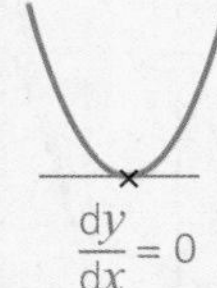

Minimum turning point

Worked examples

Differentiating integer powers

Differentiate

i $y = x^3 + 5x^2 - 7x - 8$ ii $y = \frac{4}{x^2} - \frac{1}{x}$

Solution

i $y = x^3 + 5x^2 - 7x - 8$

$\frac{dy}{dx} = 3x^2 + 5 \times 2x^1 - 7$

$= 3x^2 + 10x - 7$

ii $y = \frac{4}{x^2} - \frac{1}{x}$

$= 4x^{-2} - x^{-1}$

$\frac{dy}{dx} = 4 \times (-2)x^{-3} - (-1)x^{-2}$

$= -8x^{-3} + x^{-2} = -\frac{8}{x^3} + \frac{1}{x^2}$

Finding the gradient of a curve at a point

Find the gradient of the curve $y = 12x - 4x^2$ at the point (3, 0).

Solution

$y = 12x - 4x^2$

Differentiate: $\frac{dy}{dx} = 12 - 4 \times 2x^1 = 12 - 8x$

Substitute the x coordinate of the point (3, 0) into the gradient function:

when $x = 3$, $\frac{dy}{dx} = 12 - 8 \times 3 = 12 - 24 = -12$

so the gradient of the curve at the point (3, 0) is -12

Remember

To differentiate: 'bring the power in front of the x, then reduce the power by 1'.

Remember

If $y = ax$ then $\frac{dy}{dx} = a$

Exam tip

It is easier to rewrite the equation using negative indices before you differentiate.

Remember: $\frac{1}{x^n} = x^{-n}$

Watch out!

Take extra care when the power is negative!

When $y = x^{-3}$

then $\frac{dy}{dx} = -3x^{-4} = \frac{3}{x^4}$.

Finding a point with a given gradient

The curve **C** has equation $y = \frac{8}{x}$

Find the coordinates of the points on **C** where the gradient of the curve is –2

Solution

Re-write the equation using negative powers: $y = 8x^{-1}$

Differentiate: $\frac{dy}{dx} = -8x^{-2} = -\frac{8}{x^2}$

The gradient is –2 means $\frac{dy}{dx} = -2$, so $-\frac{8}{x^2} = -2$

Multiply by x^2 and divide by –2: $-8 = -2x^2$, so $x^2 = 4$

Square root both sides: $x = \pm 2$

When $x = 2$, $y = \frac{8}{2} = 4$ and when $x = -2$, $y = \frac{8}{-2} = -4$

So the coordinates are (2, 4) and (–2, –4)

Finding the coordinates of stationary points

The sketch shows the curve $y = 2x^3 - 3x^2 - 12x + 5$

Find the coordinates of the points P, Q and R.

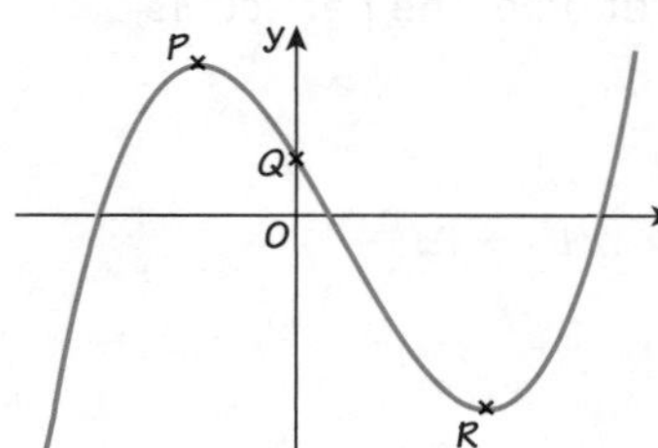

Solution

At Q: $x = 0$ so $y = 2 \times 0^3 - 3 \times 0^2 - 12 \times 0 + 5 = 5$; so $Q(0, 5)$

At P and R, $\frac{dy}{dx} = 0$ as these are stationary points.

Differentiate $y = 2x^3 - 3x^2 - 12x + 5$ to give $\frac{dy}{dx} = 6x^2 - 6x - 12 = 0$

Divide by 6: $x^2 - x - 2 = 0$

Factorise: $(x - 2)(x + 1) = 0$, and solving gives $x = 2$ or $x = -1$

When $x = -1$, $y = 2 \times (-1)^3 - 3 \times (-1)^2 - 12 \times (-1) + 5$
$= -2 - 3 + 12 + 5 = 12$

When $x = 2$, $y = 2 \times 2^3 - 3 \times 2^2 - 12 \times 2 + 5$
$= 16 - 12 - 24 + 5 = -15$

Using the graph, $P(-1, 12)$ (a maximum) and $R(2, -15)$ (a minimum).

Watch out!

Make sure you find the y coordinates – you may lose marks if you don't!

Remember

Don't forget the negative square root!

Exam tip

Use the curve to help you.

- P is at a **maximum** and has a negative x coordinate.
- R is at a **minimum** and has a positive x coordinate.

Exam tip

To find the coordinates of a **stationary or turning point**:

- **differentiate**
- **solve** $\frac{dy}{dx} = 0$ to find the x coordinates
- **substitute** the x coordinates into the **equation of the curve** to find the y coordinates.

Exam-style questions

1 The curve **C** has equation $y = 8x^2 + 3x + \frac{2}{x} - 5$

Find the coordinates of the point on **C** where the gradient of the curve is 3 [4]

2 The curve with equation $y = \frac{1}{x^2} + 2x$ has a stationary point at P.

a Work out the coordinates of P. [3]

b Work out the gradient of the curve at the point where $x = -2$ [2]

Short answers on page 138

Full worked solutions online

CHECKED ANSWERS ONLINE

Applications of differentiation

REVISED

Key facts

1. You can use differentiation with other variables to solve problems.
2. - Displacement, s, is the distance between two fixed points. s is written as a function of time.
   - Velocity, v, is the rate at which displacement changes, so $v = \frac{ds}{dt}$
   - Acceleration, a, is the rate at which velocity changes, so $a = \frac{dv}{dt}$

Examples:

$s = 5t^2 - 7t$

$v = \frac{ds}{dt} = 10t - 7$

$a = \frac{dv}{dt} = 10$

Worked examples

Solving problems involving displacement, velocity and acceleration

A particle is moving along a straight line. The fixed point O lies on this line. The displacement of the particle from O at time t seconds is s metres where $s = 4t^3 - 7t^2 + 5t - 1$

a Find an expression for the velocity, v m/s, of the particle at time t.

b Find the time at which the acceleration of the particle is instantaneously zero.

Solution

a $s = 4t^3 - 6t^2 + 5t - 1$, $v = \frac{ds}{dt} = 12t^2 - 12t + 5$

b $a = \frac{dv}{dt} = 24t - 12$

When $a = 0$, then $24t - 12 = 0$ so $t = 0.5$ seconds.

Using differentiation to solve problems

A rectangular playground has area A m^2 and perimeter 240 m.

The width of the playground is w m and the length is l m.

a Show that $A = 120w - w^2$

b Hence find the maximum area of the playground.

Solution

a Perimeter $= 2w + 2l = 240$ so $w + l = 120$ and $l = 120 - w$

Area $= lw$, so Area $= (120 - w) \times w = 120w - w^2$

b $\frac{dA}{dw} = 120 - 2w$

At a maximum, $\frac{dA}{dw} = 0$ so $120 - 2w = 0$, so $w = 60$

When $w = 60$ then $A = 120 \times 60 - 60^2 = 3600$ m^2.

Watch out!

Instantaneously means at that instant.
Instantaneously at rest means $v = 0$, but the object is still accelerating.
When the acceleration is instantaneously zero then $a = 0$, but the object may still be moving!

Remember

The units of
- displacement are m
- velocity are m/s
- acceleration are m/s^2

Exam tip

You must show all your working for 'show that' questions. Missing out steps could cost you marks!

Exam-style question

A particle is moving along a straight line. The fixed point O lies on this line.

The displacement of the particle from O at time t seconds is s metres where $s = 12t^2 + \frac{3}{t}$

a Find the displacement of the particle at time 2 seconds. [1]

b Find an expression for the velocity, v m/s, of the particle at time t. [2]

c Find the acceleration of the particle at the time when the particle is instantaneously at rest. [4]

Short answers on page 138

Full worked solutions online

CHECKED ANSWERS ONLINE

Review questions: Functions and graphs

1 The diagram shows the kite $ABCD$ where $AB = AD$ and $CB = CD$.

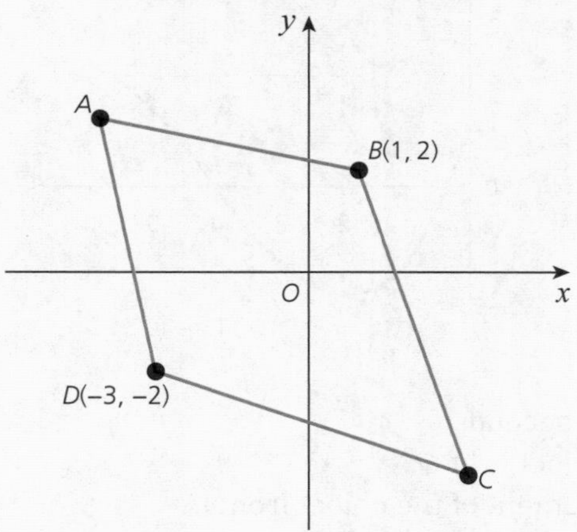

B is the point (1, 2)

D is the point (−3, −2)

The line **L** passes through the points A and C.

a Find the equation of line **L**.

Give your answer in the form $ax + by = c$ where a, b and c are integers. (4 marks)

b A is the point $(-4, p)$

C is the point $(q, -4)$

Find the value of p and the value of q. (3 marks)

2 The diagram shows the lines $\mathbf{L}_1$ and $\mathbf{L}_2$.

$\mathbf{L}_1$ passes through the points $A(-2, -1)$ and $B(8, 4)$

$\mathbf{L}_2$ is perpendicular to $\mathbf{L}_1$ and passes through the midpoint of AB.

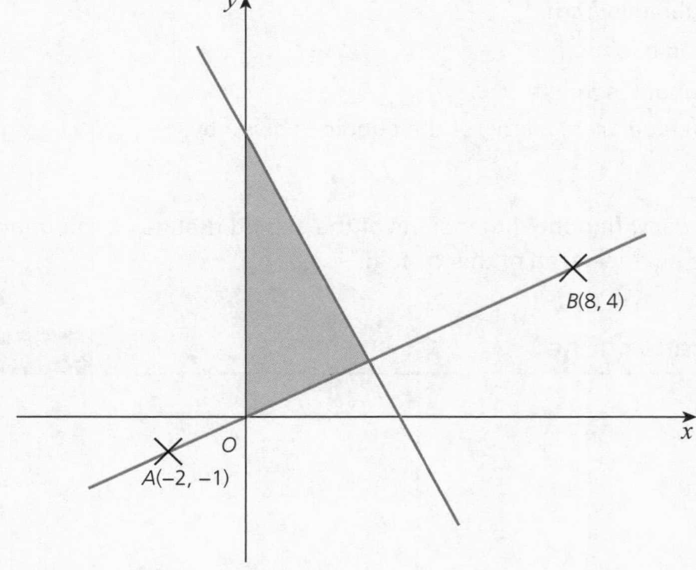

Find the area of the shaded region. (5 marks)

3 Isobel draws the graph of $y = 3x^2 - 5x + 4$

She adds the line $y = 4x - 1$ to her graph and uses it solve the equation $3x^2 + bx + c = 0$

a Find the value of b and the value of c. (3 marks)

b Find the equation of the straight line Isobel should add to her graph to solve the equation $3x^2 + 2x + 1 = 0$ (3 marks)

4 An object is moving along a straight line. The fixed point O lies on this line. The graph below shows the displacement, in metres, of the particle from O at time t seconds.

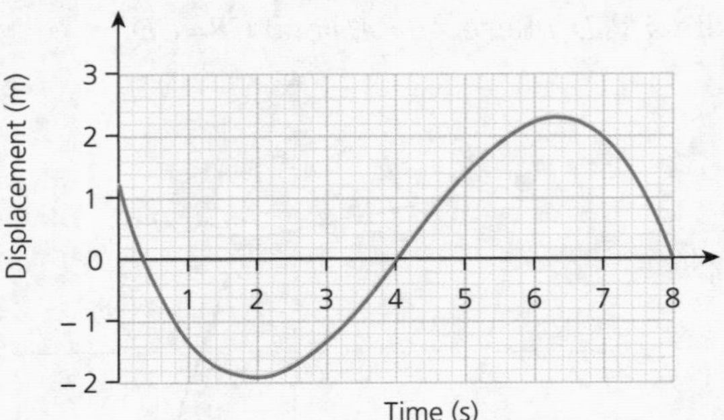

a Use the graph to find
   i the displacement at 1 second, (1 mark)
   ii the times when the object is at O, (2 marks)
   iii the maximum displacement of the object from O. (1 mark)

b The velocity of the object is positive for $a < t < b$

   Find the value of a and the value of b. (2 marks)

c Estimate the velocity at time 7 seconds. (3 marks)

5 The diagram shows a cuboid with a total volume of 288 cm^3

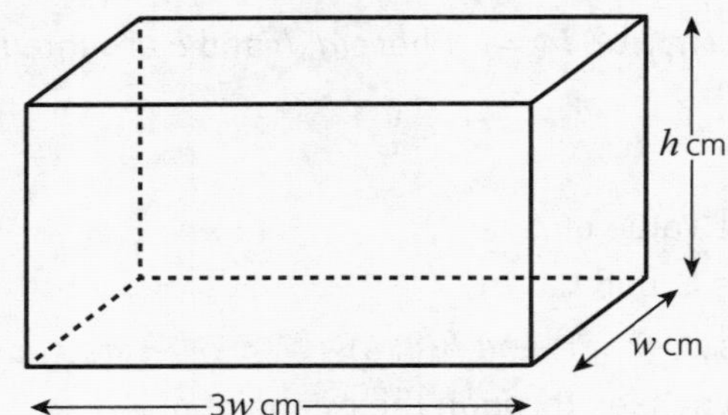

The height of the cuboid is h cm.

The width of the cuboid is w cm.

The length of the cuboid is $3w$ cm.

a Show that the surface area, A cm^2, of the cuboid is given by

$$A = 6w^2 + \frac{768}{w}$$

(3 marks)

b Given that w can vary, find the dimensions of the cuboid that has a minimum surface area and find the surface area of this cuboid. (4 marks)

Short answers on page 138

Full worked solutions online

CHECKED ANSWERS ONLINE

Target your revision: Shape, space and measure

Check how well you know each topic by answering these questions. If you struggle, go to the page number in brackets to revise the topic.

1 Find the area and perimeter of a triangle

The perimeter of triangle ABC is 21.4 cm.
Work out the area of the triangle.

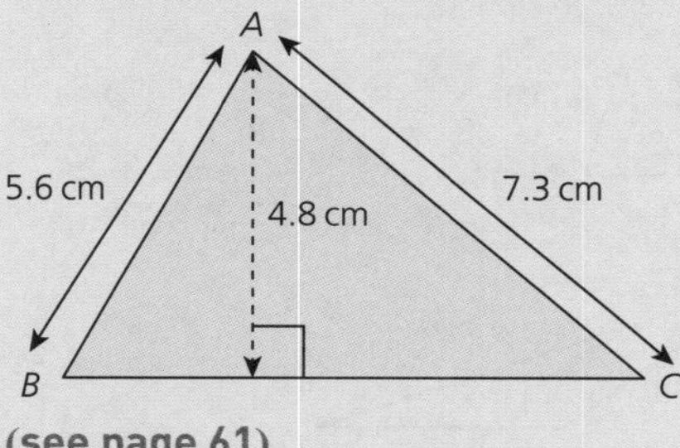

(see page 61)

2 Convert between units of area

a Convert 325 cm² to m².

b Convert 0.65 m² to mm².

(see page 61)

3 Find the area of a trapezium and parallelogram

Work out the area of these shapes.

i

ii

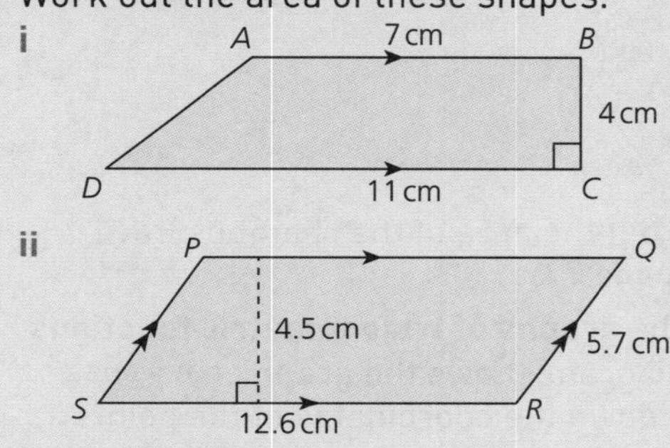

(see page 62)

4 Find the surface area of a prism

The surface area of this cuboid is 126 cm².
Work out the height of the cuboid.

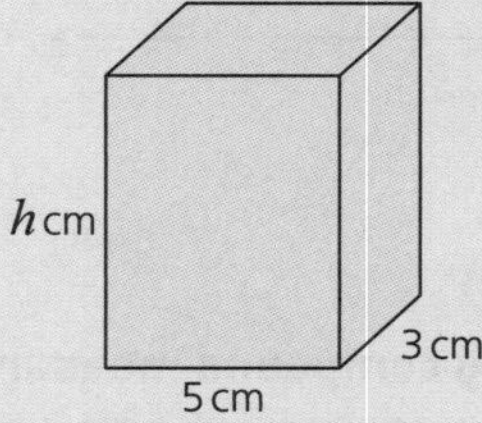

(see page 63)

5 Convert between units of volume

a Convert 123 000 cm³ to m³.

b Convert 0.315 cm³ to mm³.

(see page 64)

6 Find the volume of a prism

Calculate the volume of this prism.

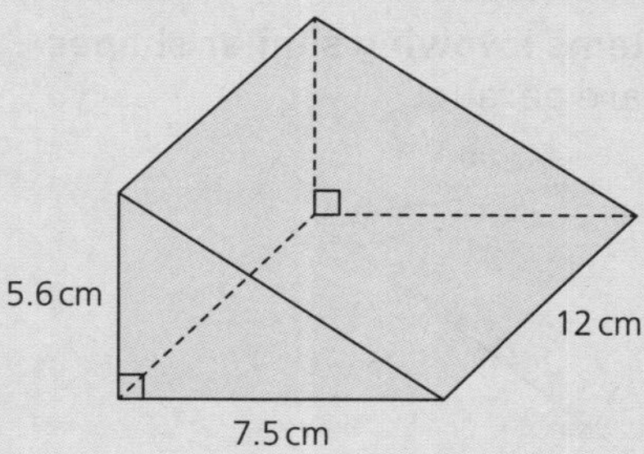

(see page 64)

7 Use the area and circumference of a circle

The diagram shows a shape made from a rectangle and a semicircle.

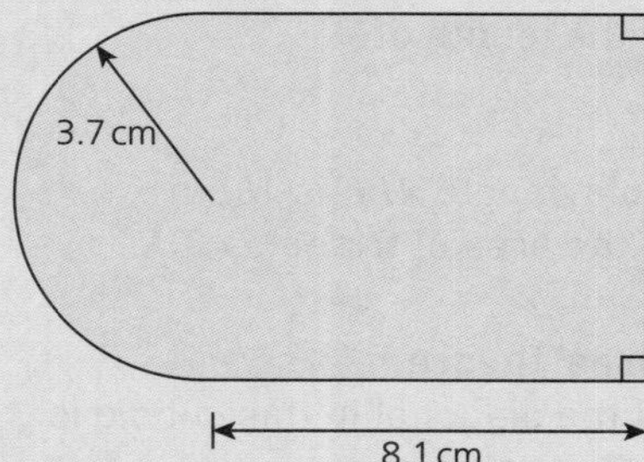

Work out the area and perimeter of the shape.

(see page 65)

8 Find the area and arc length of a sector

Find the area and perimeter of the sector AOB.

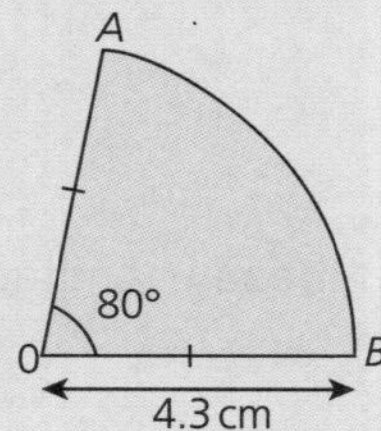

(see page 65)

9 Find the volume and surface area of cylinders and cones

A solid cone fits exactly inside a hollow cylinder of radius 6 cm and height 20 cm.

a Work out the exact volume of space not occupied by the cone.

b Work out the exact surface area of the cylinder.

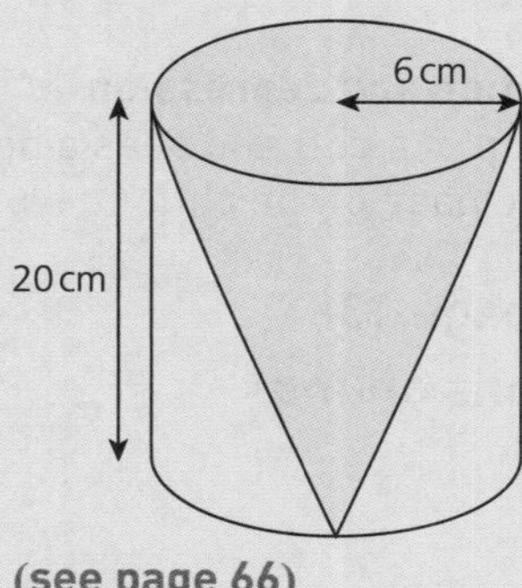

(see page 66)

10 Use the surface area and volume of a sphere

A sphere has a surface area of 81π cm². Work out the volume of the sphere.

(see page 66)

11 Solve problems involving similar shapes

AB and DE are parallel.

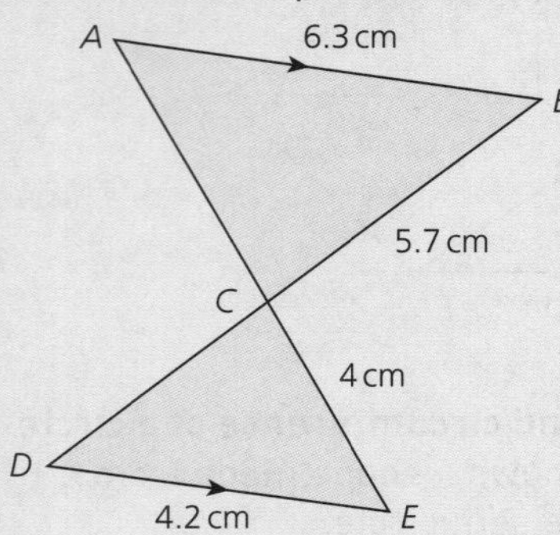

a Calculate the length of

i CD

ii AC.

b The area of triangle ABC is 16 cm². Calculate the area of triangle CDE.

(see pages 68–69)

12 Use Pythagoras' theorem

The length of the diagonal in this cuboid is $a\sqrt{a}$ cm. Find the value of a.

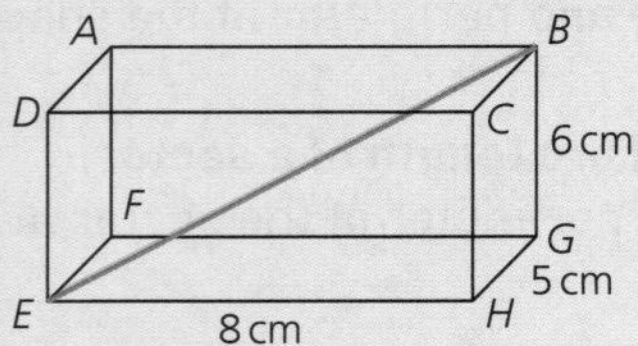

(see pages 70–71)

13 Use trigonometry

Find the value of x and of y in these triangles.

a

b

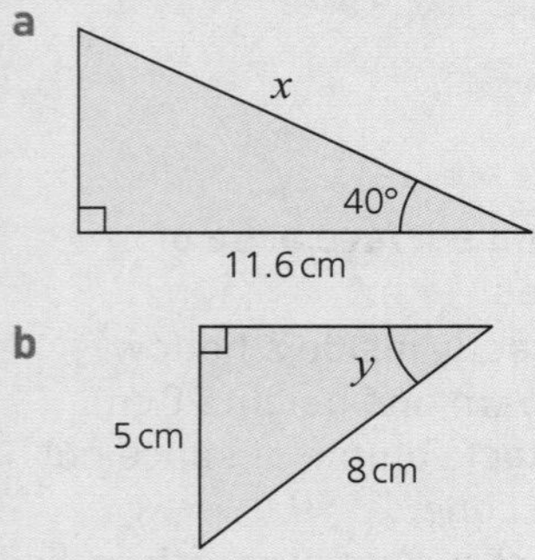

(see pages 72–73)

14 Use angles of elevation and depression

Jack stands at the top of a cliff and sees a boat 300 m away from the base of the cliff. Jack measures the angle of depression from the top of the cliff to the boat. The angle of depression is 20°. Find the height of the cliff.

(see page 74)

15 Use the sine and cosine rules

Find the value of x and of y in these triangles.

a

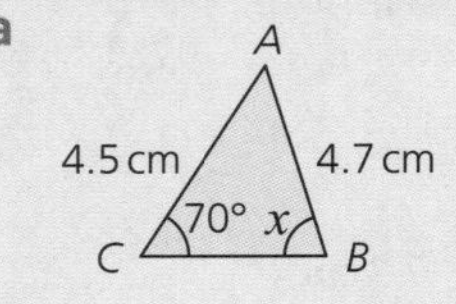

b

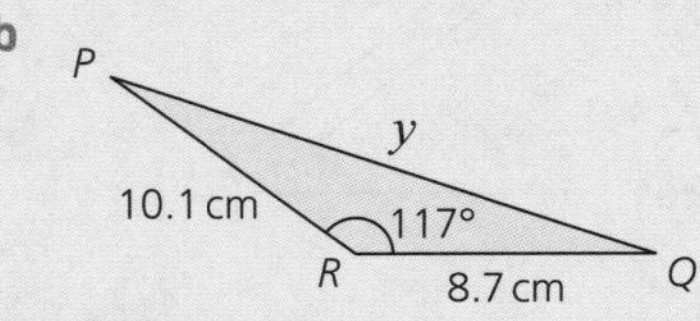

(see pages 75–76)

16 Find the area of a triangle using $\frac{1}{2}ab\sin C$

$ABCD$ is a rhombus of side 7 cm.

Angle $ABC = 100°$

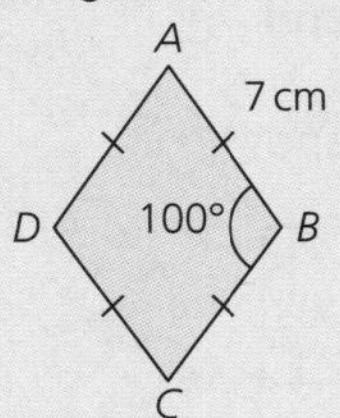

Work out the area of the rhombus $ABCD$.

(see page 77)

17 Use the graphs of trigonometric functions

The diagram shows the graph of $y = \cos x°$

Write down the coordinates of the points A, B and C.

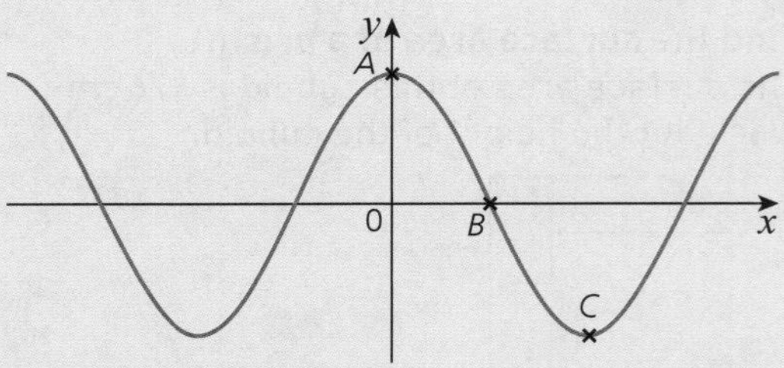

(see pages 78–79)

18 Solve problems involving compound measures

A cuboid has width 4 cm, length 5 cm and height 9 cm.

The density of the cuboid is 4.3 g/cm³

Work out the mass of the cuboid.

(see page 80)

Short answers on page 138

Full worked solutions online

CHECKED ANSWERS ONLINE

Area and perimeter

REVISED

Key facts

1 The **perimeter** of a shape is the distance around the shape.

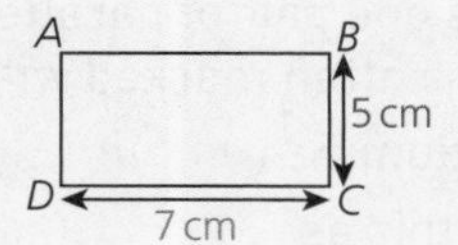

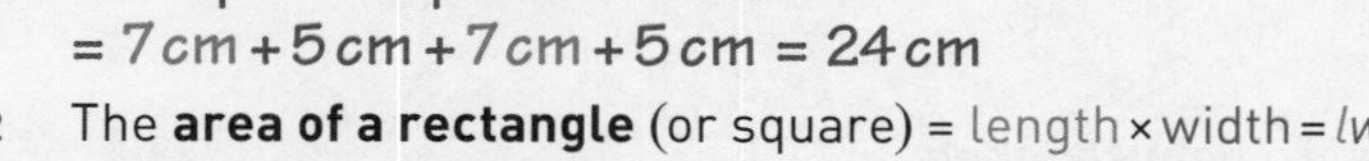

Example: The perimeter of $ABCD$ $= 7\,\text{cm} + 5\,\text{cm} + 7\,\text{cm} + 5\,\text{cm} = 24\,\text{cm}$

2 The **area of a rectangle** (or square) = length × width = lw

Example: Area $ABCD = 7\,\text{cm} \times 5\,\text{cm} = 35\,\text{cm}^2$

3 The **area of a triangle** $= \frac{1}{2}$(base×perpendicular height)

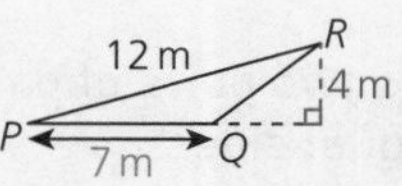

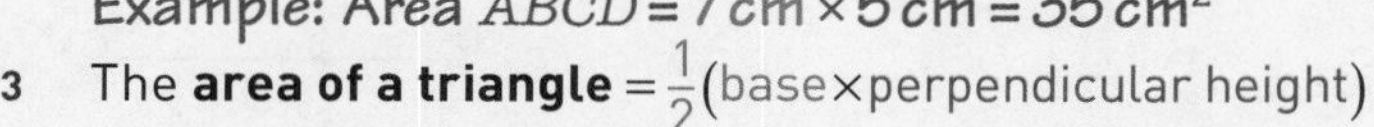

Example: Area $PQR = \frac{1}{2}(7\,\text{m} \times 4\,\text{m}) = 14\,\text{m}^2$

4 To **convert between units of area**: $1\,\text{m}^2 = 100\,\text{cm} \times 100\,\text{cm} = 10\,000\,\text{cm}^2$ and $1\,\text{cm}^2 = 10\,\text{mm} \times 10\,\text{mm} = 100\,\text{mm}^2$

Remember

1 m = 100 cm

1 cm = 10 mm

Worked examples

Finding the area of a shape

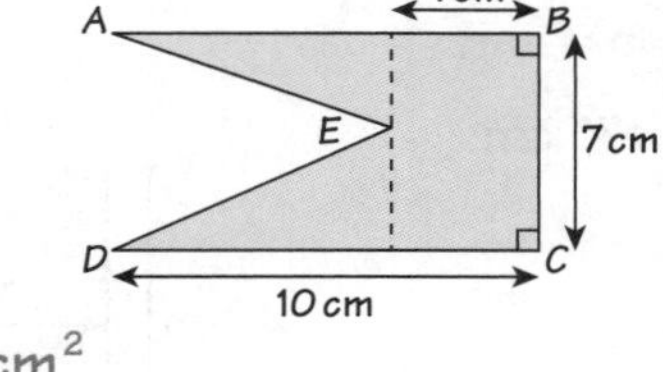

Find the area of the pentagon $ABCDE$.

Solution

Area of rectangle $ABCD = 10 \times 7 = 70\,\text{cm}^2$

Area of triangle $ADE \quad = \frac{1}{2} \times 7 \times (10 - 4) = 21\,\text{cm}^2$

Area of pentagon $ABCDE$ $70\,\text{cm}^2 - 21\,\text{cm}^2 = 49\,\text{cm}^2$

Using the perimeter of a shape

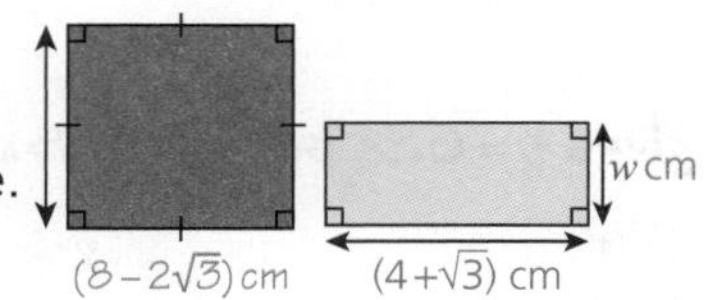

The perimeter of the square is equal to the perimeter of the rectangle.

Find the width (w cm) of the rectangle.

Solution

Perimeter of rectangle = perimeter of square

So $2(4 + \sqrt{3}) + 2w = 4(8 - 2\sqrt{3})$

Divide by 2 and expand brackets:

$4 + \sqrt{3} + w = 2(8 - 2\sqrt{3}) = 16 - 4\sqrt{3}$

Subtract $4 + \sqrt{3}$ from both sides:

$w = 16 - 4\sqrt{3} - 4 - \sqrt{3} = (12 - 5\sqrt{3})\,\text{cm}$

Convert between units of area

Convert $150.6\,\text{cm}^2$ i to mm^2 ii to m^2

Solution

i $150.6\,\text{cm}^2 = 150.6 \times 10 \times 10 = 15\,060\,\text{mm}^2$

ii $150.6\,\text{cm}^2 = 150.6 \div 100 \div 100 = 0.01506\,\text{m}^2$

Remember

Area of triangle is

$\frac{1}{2}$ base × height

or $\frac{1}{2}$ height × base

Exam tip

Don't forget the units. You may get a mark for the right units even if your calculation is wrong.

Remember

Perimeter of rectangle is $2l + 2w$

Perimeter of square of side s is $4s$

Watch out!

To convert between units of area (m^2, cm^2 or mm^2) you multiply or divide by the conversion factor twice.

Exam-style question

The perimeter of rectangle $ABCD$ is 42 cm.

a Find the area, in cm^2, of the shaded triangle PQC. [4]

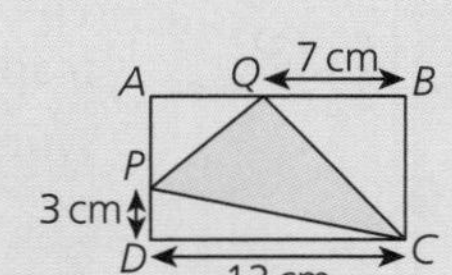

b Write your answer to part **a** in

i m^2 [1] ii mm^2 [1]

Short answers on page 139

Full worked solutions online

CHECKED ANSWERS ONLINE

Further area

REVISED

Key facts

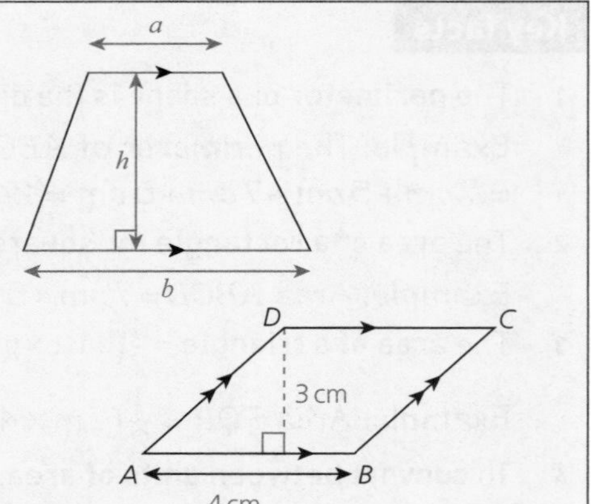

1 A **trapezium** has one pair of parallel sides.
Parallel sides are often marked with arrows.
Area of a trapezium $= \frac{1}{2}(a+b)h$
You can think of this as
'half the sum of the parallel sides × the distance between them'

2 A **parallelogram** has two pairs of parallel sides.
It has two pairs of equal sides.

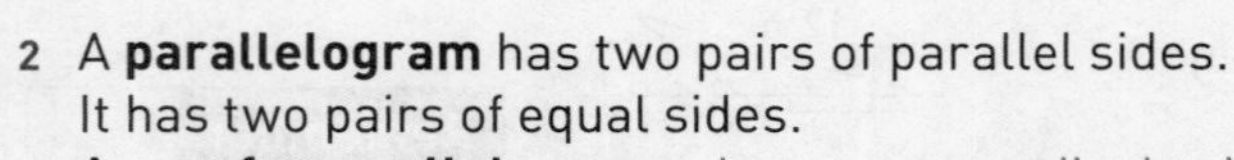

Area of a parallelogram = base × perpendicular height
Example: Area of $ABCD$ is $4 \times 3 = 12$ cm

Worked examples

Using the formula for area of a trapezium

The area of the trapezium $PQRS$ is 17.1 cm^2.

Find the length of QR.

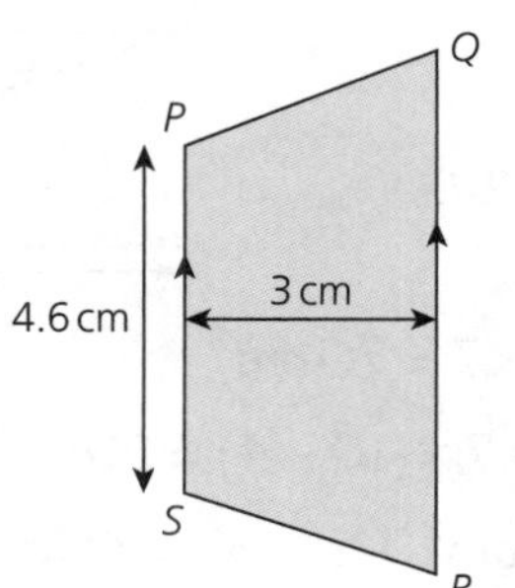

Solution

Area of trapezium $\frac{1}{2}(4.6 + b) \times 3 = 17.1$

Multiply both sides by 2 and divide by 3:

$4.6 + b = \frac{17.1 \times 2}{3} = 11.4$

Subtracting 4.6 from both sides gives $b = 6.8$; so $QR = 6.8$ cm.

Using the formula for area of a parallelogram

The area of the quadrilateral $ABCD$ is 25.08 cm^2.

Work out the area of the quadrilateral $DCFE$.

Solution

$ABCD$ is a parallelogram.
Area of a parallelogram $= DC \times 3.8$
so $DC = 25.08 \div 3.8 = 6.6$ cm

$DCFE$ is a trapezium, so the area $= \frac{1}{2}(12.4 + 6.6) \times 4.2$
$= 39.9$ cm^2

Watch out!

The 'height' of the trapezium is perpendicular to the parallel sides.

Exam tip

Always check the formula sheet – you will find the formula for the area of a trapezium there.

Remember

A **quadrilateral** is any four-sided shape. Rectangles, squares, parallelograms and trapezia are special types of quadrilateral.

Exam-style questions

1 The perimeter of parallelogram $DEFG$ is 29 cm.
Find the area of the parallelogram. [3]

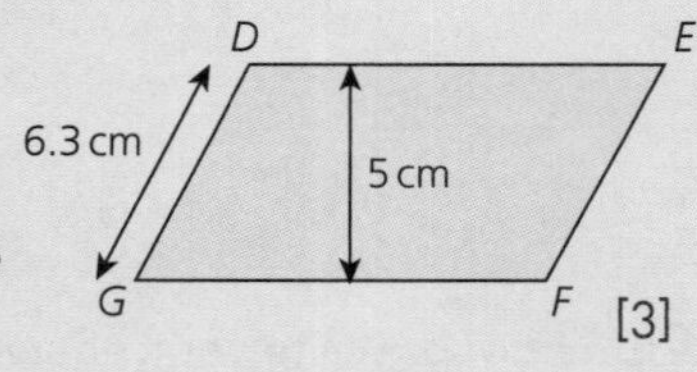

2 $ABCD$ is a trapezium.
All measurements are in metres.
The area of the trapezium is 40 m^2.
Work out the exact value of x. [4]

Short answers on page 139
Full worked solutions online

CHECKED ANSWERS ONLINE

Surface area

REVISED

Key facts

1 A **face** is a flat surface of a 3-dimensional shape.
An **edge** is formed when two faces meet.
A **vertex** is a corner where several faces or edges meet.
Example: A cuboid has 6 faces, 12 edges and 8 vertices

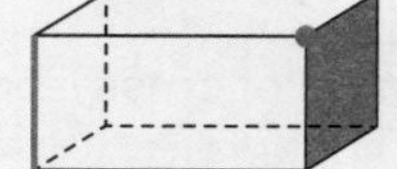

2 The **surface area** of a 3-dimensional shape is the sum of the areas of its faces.
Example: Surface area of a cuboid is:

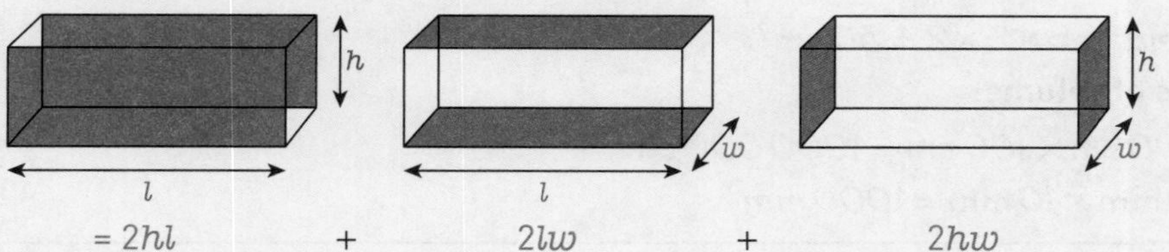

$= 2hl \quad + \quad 2lw \quad + \quad 2hw$

3 A **prism** is a 3-dimensional shape with a constant cross-section.
It has two identical end faces and flat sides.
The sides are parallelograms or rectangles.

Worked example

Find the surface area of a prism

The diagram shows a triangular prism.
The surface area of the prism is 138 cm².
Work out the length of the prism.

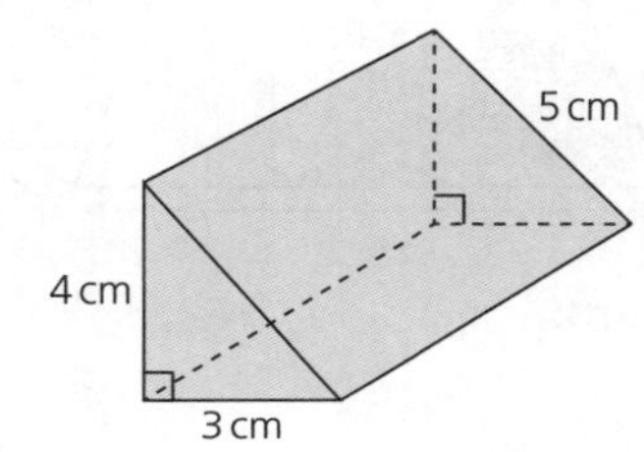

Solution

The surface area of the prism is the area of

2 × triangular end faces + base + sloping side + vertical side

So $2 \times \frac{1}{2}(3 \times 4) + 3 \times \text{length} + 5 \times \text{length} + 4 \times \text{length} = 138$

Let the length of the prism $= x$

So $12 + 3x + 5x + 4x = 138$

Simplifying gives $12 + 12x = 138$; so $12x = 126$ and $x = 10.5$

The length of the prism is 10.5 cm

Watch out!

Count the number of faces before you start so you don't miss any.

Remember

The other sides are rectangles.

Exam tip

Read the question carefully, don't muddle up surface area and volume.

Exam-style question

The diagram shows a cube and a square-based pyramid.

The surface area of the cube is the same as the surface area of the pyramid.

Work out the value of x.

Give your answer correct to 3 significant figures. [4]

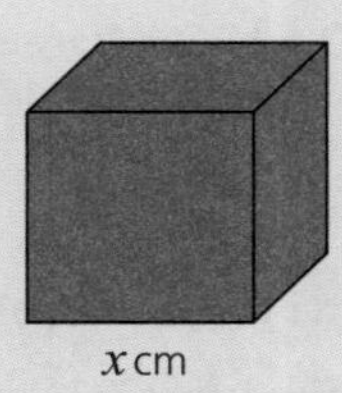

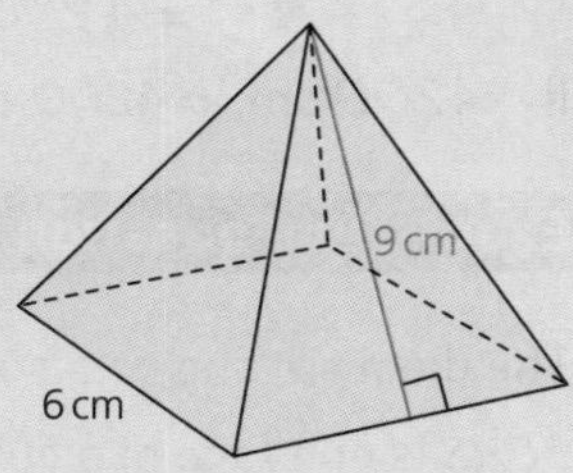

Short answers on page 139
Full worked solutions online

CHECKED ANSWERS ONLINE

Volume

REVISED

Key facts

1 The **volume of a prism** is area of cross-section × length.

Example:

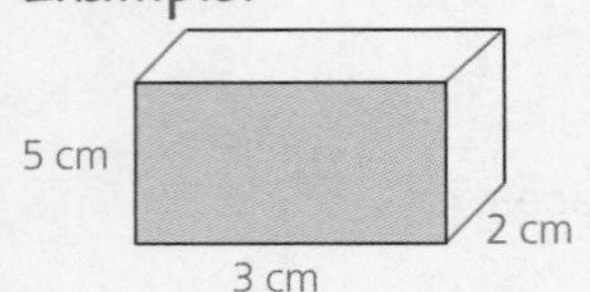

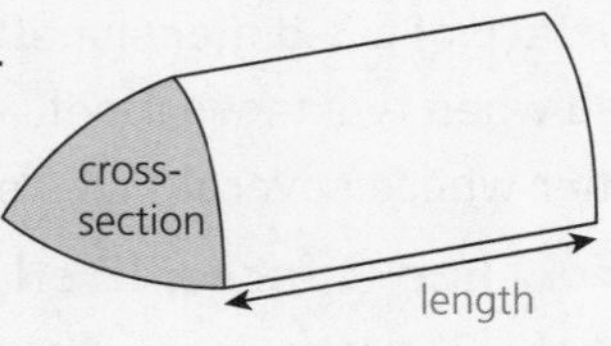

The volume of the cuboid is $5 \times 3 \times 2 = 30\,\text{cm}^3$.

2 To **convert between units of volume**:

$1\,\text{m}^3 = 100\,\text{cm} \times 100\,\text{cm} \times 100\,\text{cm} = 1\,000\,000\,\text{cm}^3$

and $1\,\text{cm}^3 = 10\,\text{mm} \times 10\,\text{mm} \times 10\,\text{mm} = 1000\,\text{mm}^3$

Worked examples

Volume of a prism

The diagram shows a solid prism.

Work out the volume of the prism.

Give your answer correct to 3 significant figures.

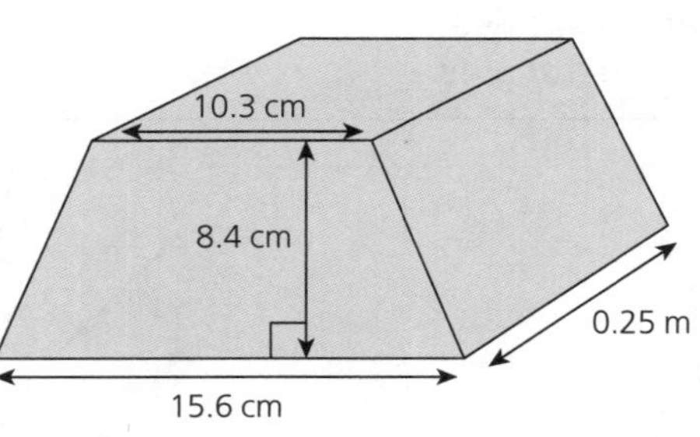

Solution

The cross-section is a trapezium.

Area of a trapezium $= \frac{1}{2}(a + b)h$

$= \frac{1}{2}(15.6 + 10.3) \times 8.4 = \frac{1}{2} \times 25.9 \times 8.4 = 108.78\,\text{cm}^2$

Length of prism $= 0.25\,\text{m} = 25\,\text{cm}$

Volume of prism $= 108.78 \times 25 = 2719.5\,\text{cm}^3$

$= 2720\,\text{cm}^3$ correct to 3 s.f.

Convert between units of volume

Convert $4200\,\text{cm}^3$ to

i mm^3 ii m^3

Give your answers in standard form.

Solution

i $4200\,\text{cm}^3 = 4200 \times 10 \times 10 \times 10$

$= 4\,200\,000\,\text{mm}^3 = 4.2 \times 10^6\,\text{mm}^3$

ii $4200\,\text{cm}^3 = 4200 \div 100 \div 100 \div 100 = 0.0042\,\text{m}^3 = 4.2 \times 10^{-3}\,\text{m}^3$

Exam tip

Don't round until you reach the final answer – otherwise you may lose marks.

Watch out!

Make sure you are using the same units for each of the 'sides'.

Remember

See page 12 for a reminder of how to use standard form.

Exam-style question

The diagram shows a solid prism with cross section *PSRUTQ*.

PQRS is one face of a cube of volume $216\,\text{cm}^3$.

RQTU is a parallelogram.

The length of the prism is 15 cm.

Work out the total volume of the prism. [4]

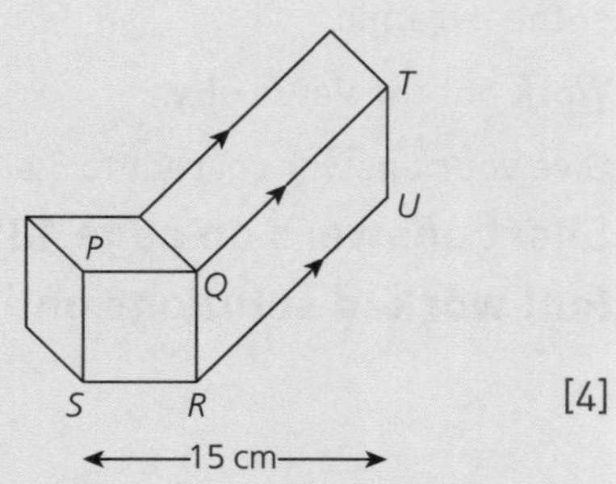

Short answers on page 139

Full worked solutions online

CHECKED ANSWERS ONLINE

Circles

REVISED

Key facts

1 The **circumference**, C, of a circle is the distance all the way round it.
$C = \pi d$ where d = diameter and $\pi = 3.141592...$
Or $C = 2\pi r$ where r = radius.
Example: The circumference of a circle of diameter 8 cm is $\pi \times 8 = 25.1$ cm to 3 s.f.

diameter
radius

2 The **area**, A, of a circle is $A = \pi r^2$
Example: The area of a circle of diameter 8 cm (so radius 4 cm) is $\pi \times 4^2 = 50.3\text{ cm}^2$ to 3 s.f.

3 A **sector** of a circle looks like a slice of pie.
A **chord** divides a circle into two **segments**.
The area of a sector of angle θ is $\frac{\theta}{360} \times \pi r^2$
The arc length of a sector of angle θ is $\frac{\theta}{360} \times 2\pi r$

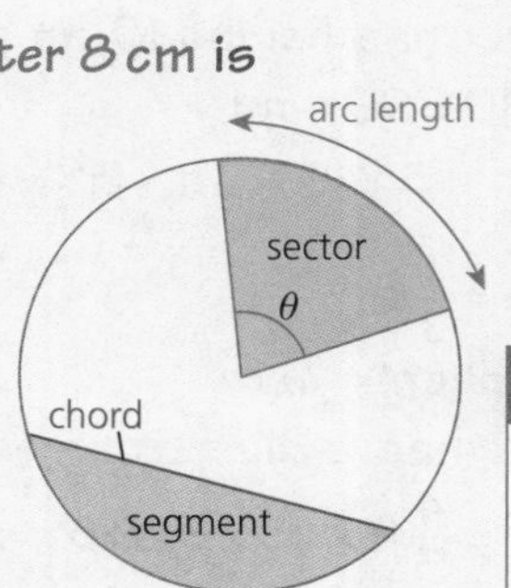

Remember

See page 77 for an example of finding the area of a segment.

Worked examples

Find the area and circumference of a circle

Find the exact area and perimeter of the shaded regions between the square and the circle.

10 mm

Solution

Shaded area = area of square − area of circle
$= 10^2 - \pi \times 5^2 = (100 - 25\pi)\text{ mm}^2$

Perimeter = perimeter of square + circumference of circle
$= 4 \times 10 + \pi \times 10 = (40 + 10\pi)\text{ mm}$

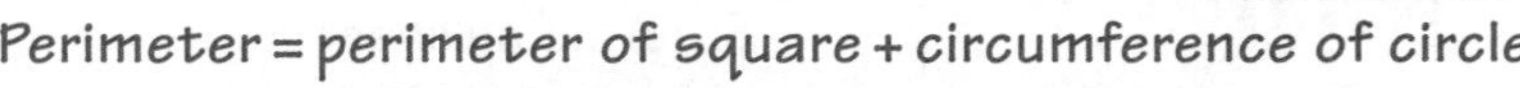

Finding the area and perimeter of a sector

Find the area and perimeter of this sector.

7 m
120°

Solution

Area of sector $= \frac{120}{360} \times \pi \times 7^2 = 51.3\text{ m}^2$ to 3 s.f.

Arc length $= \frac{120}{360} \times 2\pi \times 7 = 14.66...\text{ m}$

Perimeter $= 14.66... + 2 \times 7 = 28.7$ m to 3 s.f.

Watch out!

Leave π in your answer if you are asked for an exact answer.

Exam tip

Use your calculator key for π.

Remember

A sector is a fraction of a whole circle.
In this example, $\frac{120}{360} = \frac{1}{3}$, so the sector is $\frac{1}{3}$ of the whole circle.

Exam-style questions

1 The shaded shape is made by cutting four identical quarter circles from a rectangular piece of card.
Work out the area and perimeter of the shaded shape.
Give your answer correct to 3 significant figures. [4]

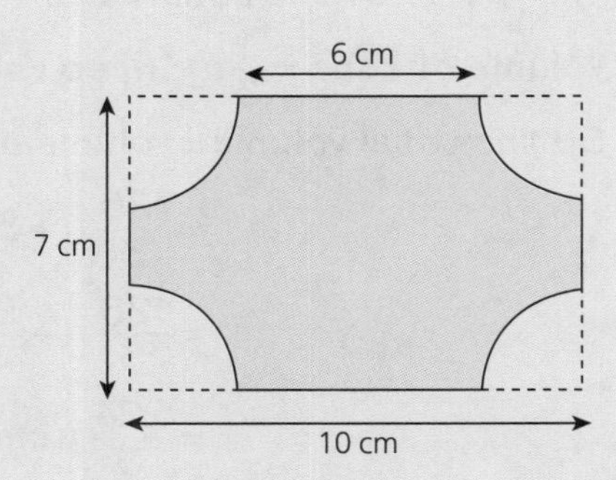

2 A sector of a circle of radius 6 cm has area $25\pi\text{ cm}^2$.
Work out the perimeter of the sector.
Give your answer in terms of π. [5]

Short answers on page 139
Full worked solutions online

CHECKED ANSWERS ONLINE

Cylinders, spheres and cones

REVISED

Key facts

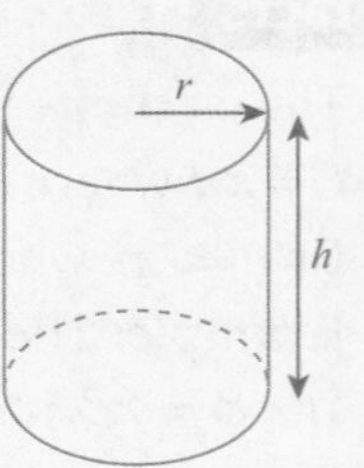

1 **Volume of a cylinder** $= \pi r^2 h$

Surface area of cylinder = area of curved surface + area of two ends

$= 2\pi rh + 2\pi r^2$

Example: A cylinder has height 10 cm and radius 3 cm

Volume $= \pi \times 3^2 \times 10 = 90\pi\,\text{cm}^3$

Surface area $= 2\pi \times 3 \times 10 + 2\pi \times 3^2 = 60\pi + 18\pi = 78\pi\,\text{cm}^2$

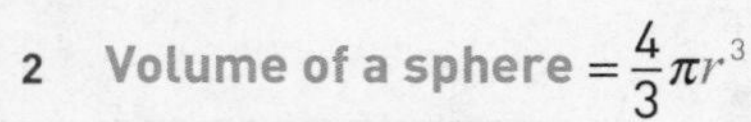

2 **Volume of a sphere** $= \frac{4}{3}\pi r^3$

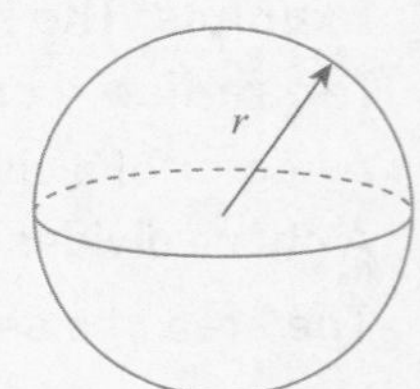

Surface area of a sphere $= 4\pi r^2$

Example: A sphere has radius 6 cm

Volume $= \frac{4}{3}\pi \times 6^3 = \frac{4}{3}\pi \times 216 = 288\pi\,\text{cm}^3$

Surface area $= 4\pi \times 6^2 = 144\pi\,\text{cm}^2$

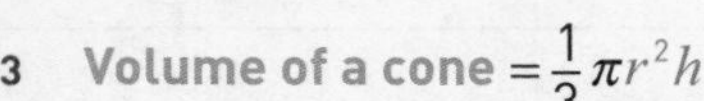

3 **Volume of a cone** $= \frac{1}{3}\pi r^2 h$

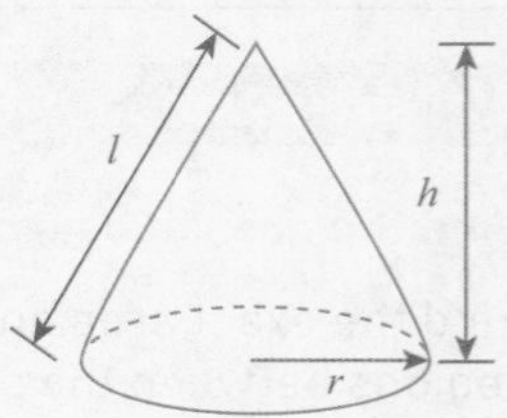

Surface area of a cone = area of curved surface + area of base $= \pi rl + \pi r^2$

Example: A cone has height 8 cm, slant height 10 cm and radius 6 cm

Volume $= \frac{1}{3}\pi \times 6^2 \times 8 = 96\pi\,\text{cm}^3$

Surface area $= \pi \times 6 \times 10 + \pi \times 6^2 = 60\pi + 36\pi = 96\pi\,\text{cm}^2$

Worked examples

Finding the volume of a solid

A solid shape consists of a hemisphere of radius 4 cm attached to a cone.

The total height of the solid is 10 cm.

Work out the exact volume of the solid.

Solution

Volume of hemisphere $= \frac{1}{2}$ volume of sphere

$$= \frac{2}{3}\pi r^3$$

When $r = 4$ cm, volume of hemisphere is

$$\frac{2}{3}\pi \times 4^3 = \frac{2}{3}\pi \times 64 = \frac{128}{3}\pi\,\text{cm}^3$$

The height of the cone = 10 − 4 = 6 cm

The radius of the cone = 4 cm

Volume of cone $= \frac{1}{3}\pi r^2 h$, so volume $= \frac{1}{3}\pi \times 4^2 \times 6 = 32\pi\,\text{cm}^3$

So the total volume = volume of hemisphere + volume of cone

$$= \frac{128}{3}\pi + 32\pi$$

$$= \frac{128}{3}\pi + \frac{96}{3}\pi$$

$$= \frac{224}{3}\pi\,\text{cm}^3$$

Exam tip

Don't forget to use your formulae sheet at the front of your exam paper. You don't need to learn these formulae, but you do need to be confident in using them!

Watch out!

You were asked to leave your answer in terms of π, so don't give your answer as a decimal.

Finding the surface area of a solid

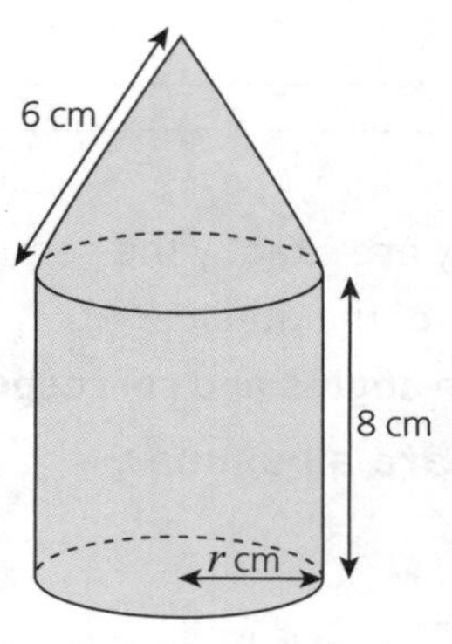

The diagram shows a solid shape made from a cone on top of a cylinder.

The cone has a radius of r cm and a height of 8 cm.

The cylinder has a slant height of 6 cm.

The total surface area of the solid is $75\pi \text{cm}^2$.

Work out the radius of the cylinder.

Solution

The total surface area = curved surface of cone + curved surface of cylinder + circular base

So total surface area $= \pi rl + 2\pi rh + \pi r^2$

$l = 6$, $h = 8$ and total surface area $= 75\pi$

So $\pi r \times 6 + 2\pi r \times 8 + \pi r^2 = 6\pi r + 16\pi r + \pi r^2 = 75\pi$

Simplify: $22\pi r + \pi r^2 = 75\pi$

Divide by π: $22r + r^2 = 75$

This is a quadratic in r, so $r^2 + 22r - 75 = 0$

Factorise: $(r + 25)(r - 3) = 0$

So $r = -25$ (impossible as r is a length and so has to be positive) or $r = 3$

So the radius of the cylinder is 3 cm.

Watch out!

The diagram will not be drawn to scale so don't try and measure the lengths.

Remember

Don't forget to include the base of the solid when you find its surface area.

Exam tip

Make sure you show the examiner you are disregarding any 'impossible' solutions. You may lose a mark if the examiner thinks you mean to leave –25 as an answer.

Exam-style questions

1 The diagram shows a solid cone.

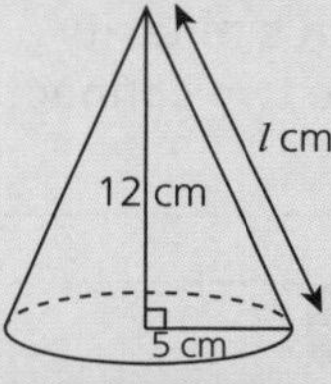

Remember

r, h and l form a right-angled triangle.
See page 70 for a reminder of Pythagoras' theorem.

a Work out the volume of the cone. [3]

b Work out the total surface area of the cone. [3]

2 The diagram shows a solid sphere and a solid cylinder.

The volume of the sphere is the same as the volume of cylinder.

The radius of the sphere is the same as the radius of the cylinder.

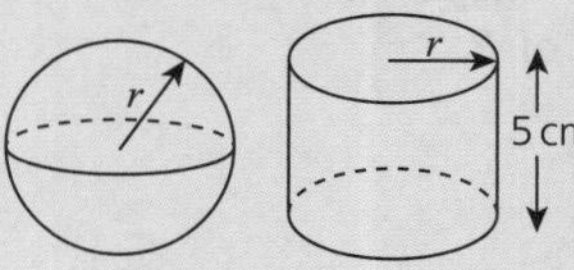

The height of the cylinder is 5 cm.

a Work out the radius of the sphere. [4]

b Show that the surface area of the cylinder is $\frac{75}{8}\pi \text{cm}^2$ greater than the surface area of the sphere. [4]

3 A ball is packed into a cube-shaped box.

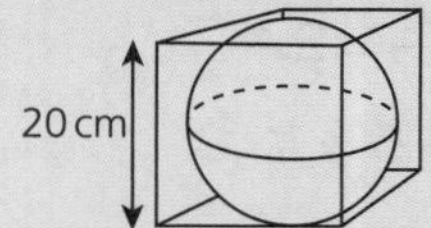

The box is 20 cm high.

The ball is a perfect sphere.

Work out the volume of space in the cube **not** occupied by the ball. [4]

Short answers on page 139

Full worked solutions online

CHECKED ANSWERS ONLINE

Similarity

REVISED

Key facts

1 Two objects are **similar** if they are exactly the same shape but have different sizes. One shape is an **enlargement** of the other. Similar shapes have the **same angles** and **corresponding sides** are in the **same ratio**.

Example: These triangles are all similar.

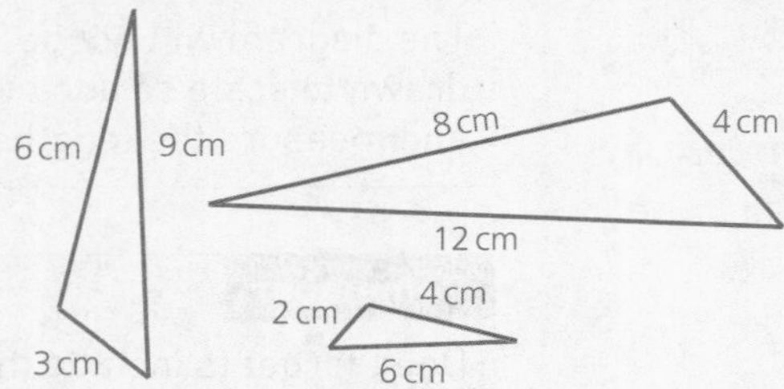

Remember

Similar shapes may be flipped (reflected) or turned (rotated) as well as enlarged.

2 When the scale factor of the enlargement is n then corresponding:

- lengths are in the ratio $1:n$
- areas are in the ratio $1:n^2$
- volumes are in the ratio $1:n^3$

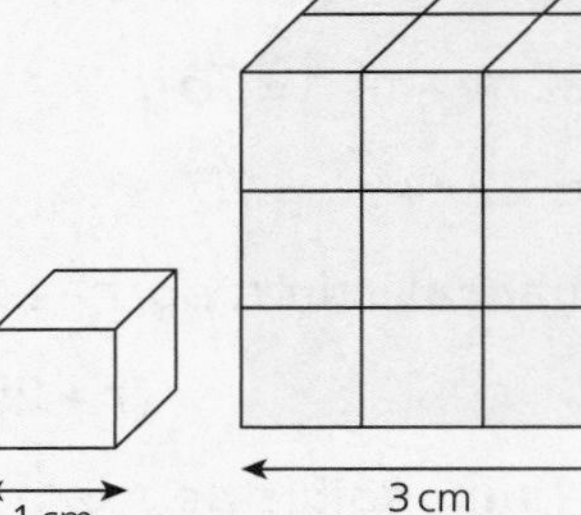

Example: These cubes are similar.

The cubes have

- side lengths in ratio 1:3
- surface areas in ratio 1:9
- volumes in ratio 1:27

Worked examples

Finding lengths in similar shapes

TS and PQ are parallel.

a Calculate the length of RQ.

b Calculate the length of RS.

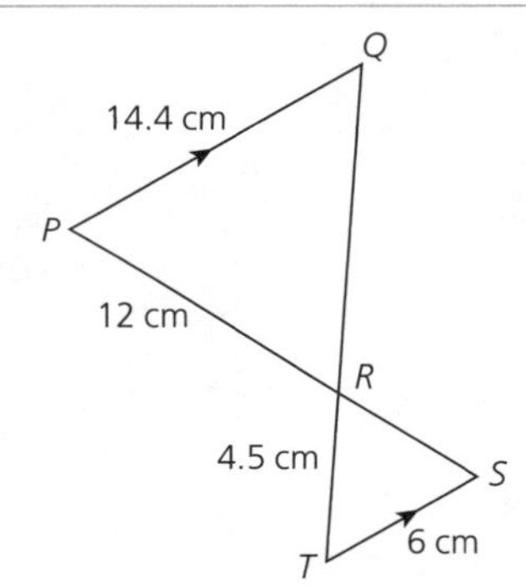

Solution

The two triangles are similar as they have the same angles.

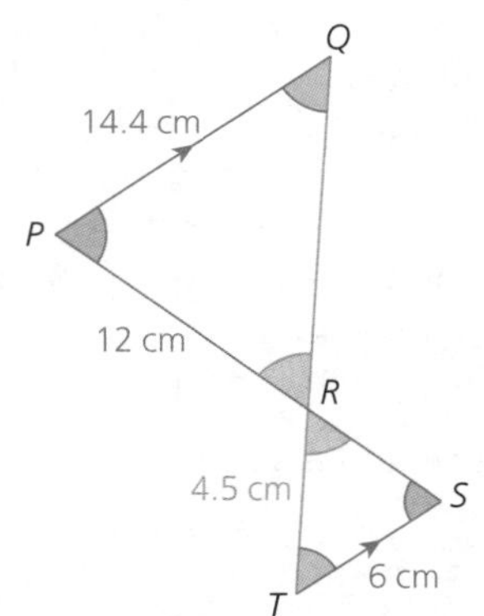

Find the scale factor, n, first:

Scale factor $n = \frac{PQ}{TS} = \frac{14.4}{6} = 2.4$

a $RQ = 2.4 \times TR = 2.4 \times 4.5 = 10.8$

b $RS = PR \div 2.4 = 12 \div 2.4 = 5$

Exam tip

Questions on similarity often involve two connected triangles.

Remember

See page 86 for a reminder of alternate and vertically opposite angles. The triangles have the same angles – so they are similar.

Watch out!

The shapes may not be the same way around – make sure you match up the corresponding sides correctly.

Finding areas in similar shapes

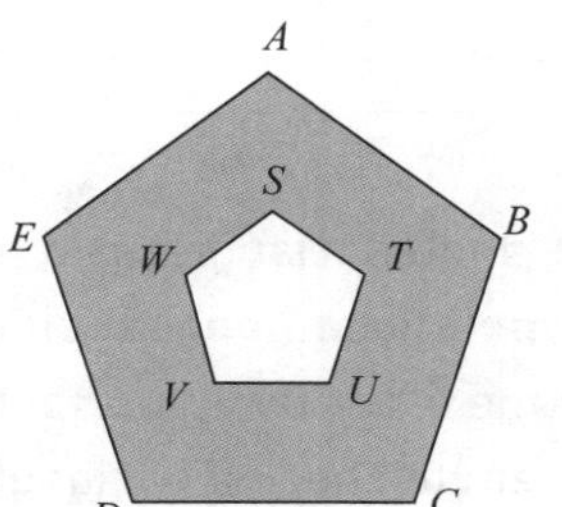

The diagram shows two regular pentagons, $ABCDE$ and $STUVW$.

$AB = 5.5$ cm and $ST = 2.2$ cm.

The area of $ABCDE = 52\text{ cm}^2$.

Find the area of the shaded region.

Solution

First find the scale factor, n:

Scale factor $n = \frac{AB}{ST} = \frac{5.5}{2.2} = 2.5$

Area of $ABCDE = n^2 \times$ area of $STUVW$

So $52 = 2.5^2 \times$ area of $STUVW$

So the area of $STUVW = \frac{52}{2.5^2} = 8.32\text{ cm}^2$

Shaded area $= 52 - 8.32 = 43.68\text{ cm}^2$

Finding volumes in similar shapes

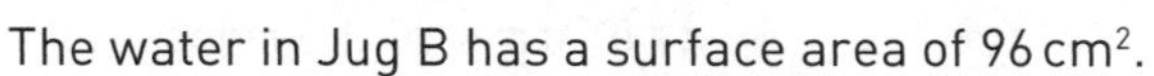

Jug A   Jug B

The diagram shows two mathematically similar jugs of water, A and B.

The water in Jug A has a surface area of 150 cm^2 and a volume of 1250 cm^3.

The water in Jug B has a surface area of 96 cm^2.

Work out the volume of water in Jug B.

Solution

First find the scale factor, n:

$$n^2 = \frac{\text{surface area of Jug A}}{\text{surface area of Jug B}} = \frac{150}{96} = \frac{25}{16}$$

So $n = \sqrt{\frac{25}{16}} = \frac{5}{4} = 1.25$

Volume of water in Jug A $= n^3 \times$ volume of water in Jug B

So volume of water in Jug B $=$ volume of water in Jug A $\div n^3$

$= 1250 \div 1.25^3$

$= 640\text{ cm}^3$

Exam tip

Any two regular shapes with the same number of sides are similar. So all equilateral triangles are similar, likewise all squares are similar, all regular pentagons are similar and so on.

Watch out!

Make sure you answer the question! You had to find the shaded area, not just the area of $STUVW$.

Remember

If you double the side lengths then the area becomes 4 times greater and the volume is 8 times greater!
So the scale factor for the volume is cubed.

Exam-style questions

1 In the diagram, AED and BCD are straight lines.

AB is parallel to EC.

$AE = 4.8$ cm, $ED = 6$ cm, $EC = 10.5$ cm, $CD = 8$ cm.

A, B, C, D, E; 10.5 cm; 4.8 cm; 8 cm; 6 cm

a Work out the length of AB. [2]

b Work out the length of BC. [2]

The area of triangle CDE is 23.86 cm^2.

c Work out the area of the quadrilateral $ABCE$. [3]

2 A and B are two mathematically similar prisms.

Prism A has a length of 20 cm and a volume of 250 cm^3.

Prism B has a volume of 1024 cm^3. Work out the length of Prism B. [3]

Short answers on page 139

Full worked solutions online

CHECKED ANSWERS ONLINE

Pythagoras' theorem

REVISED

Key facts

1. Pythagoras' theorem for right-angled triangles is $c^2 = a^2 + b^2$
   Where c is the length of the **hypotenuse (longest side)**
   and a and b are the lengths of the other two (shorter) sides.
2. You can use Pythagoras' theorem to check if a triangle is right-angled.
   When the 3 sides of a triangle satisfy the equation $c^2 = a^2 + b^2$ then the triangle is right-angled.

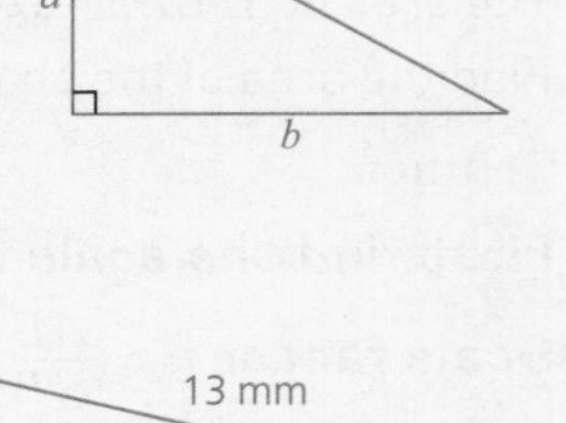

Example: A triangle with sides 5 mm, 12 mm and 13 mm is right-angled as $13^2 = 5^2 + 12^2$
So $169 = 25 + 144$ ✓

Worked examples

Finding the length of the hypotenuse

Work out the length of AC. Give your answer correct to 3 significant figures.

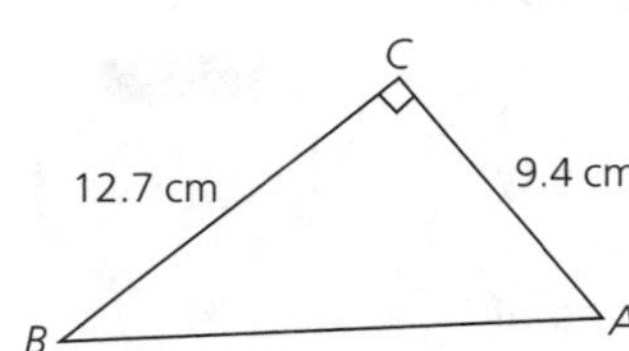

Solution

Using $c^2 = a^2 + b^2$ with $a = 12.7$, $b = 9.4$ and $c = AB$ gives:

$c^2 = 12.7^2 + 9.4^2 = 161.29 + 88.36 = 249.65$

$c = \sqrt{249.65} = 15.80\ldots$

So $AB = 15.8$ cm correct to 3 s.f.

Finding the length of a shorter side

The diagram shows triangle PQR.
Work out the area of the triangle.

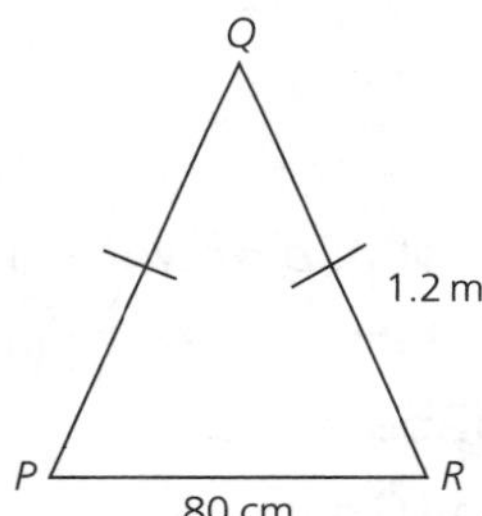

Solution

First find the height of the triangle:

$\frac{1}{2} PR = 40 \text{ cm} = 0.4 \text{ m}$

Using $a^2 + b^2 = c^2$ with $a = 0.4$, $b = h$ and $c = 1.2$ gives:

$0.4^2 + h^2 = 1.2^2$

Simplifying gives: $h^2 = 1.2^2 - 0.4^2 = 1.28$

Square root both sides: $h = \sqrt{1.28}$

Area of triangle $= \frac{1}{2} \times 0.8 \times \sqrt{1.28}$
$= 0.4525\ldots$
$= 0.453 \text{m}^2$ to 3 s.f.

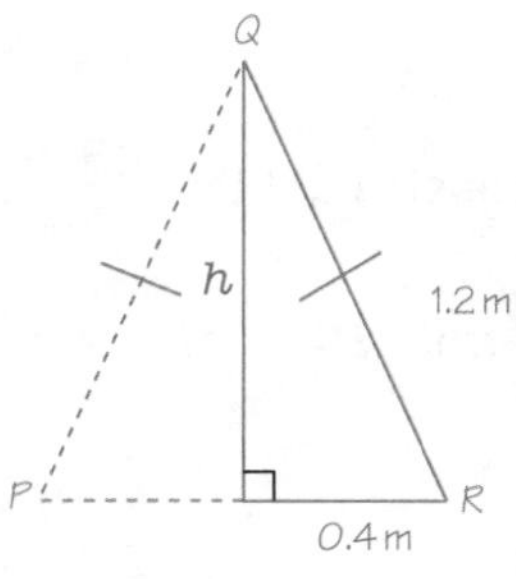

Exam tip

The longest side is opposite the right angle.

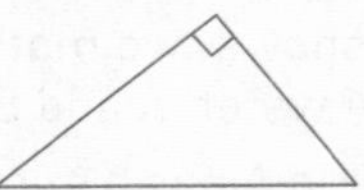

The shorter sides are the two 'arms' of the right-angle.

Remember

To find the longest side:
- Square both sides
- ADD
- Square root

Watch out!

The sides must have the same units.

Remember

To find one of the shorter sides:
- Square both sides
- SUBTRACT
- Square root

Remember

It doesn't matter which of the shorter sides you call 'a' and which 'b'.

Finding the distance between two points

The point P is at (–3, –2) and the point Q is at (1, 3). Find the distance PQ.

Solution

Difference in x-coordinates is 4
Difference in y-coordinates is 5
Using Pythagoras' theorem:

$PQ^2 = 4^2 + 5^2 = 16 + 25 = 41$

$PQ = \sqrt{41} = 6.40$ units

Remember, in general:

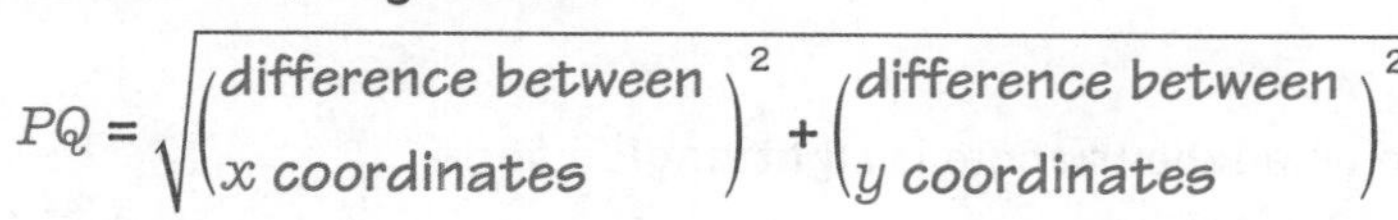

Using Pythagoras' theorem in 3-dimensions

$ABCD$ is the square base of pyramid $VABCD$.
$AB = 6$ m and $VA = VB = VC = VD = 10$ m
Calculate the height of the pyramid.
Give your answer correct to 3 significant figures.

Solution

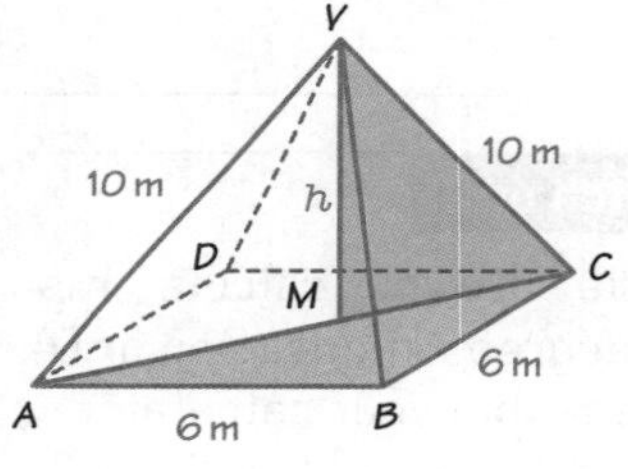

$AC^2 = 6^2 + 6^2 = 72$

$AC = \sqrt{72}$

$h^2 = 10^2 - \left(\frac{1}{2}\sqrt{72}\right)^2 = 100 - \frac{1}{4} \times 72 = 82$

$h = \sqrt{82} = 9.055... = 9.06$ m to 3 s.f.

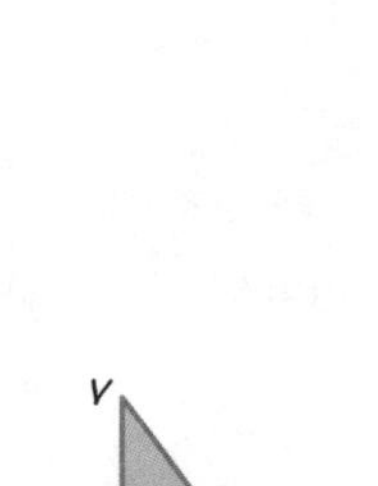

Exam tip

Draw a diagram to help you.

Remember

Pythagoras' theorem only works for **right-angled triangles**.
Use Pythagoras' theorem when you know two sides of a right-angled triangle and you want to find the third side.
For questions involving triangles and angles you need to use trigonometry – see page 72.

Watch out!

Take care when identifying the right-angled triangles. Don't forget that the angles look distorted on a 3-dimensional diagram. For example, angle ABC is a right angle as $ABCD$ is a square.

Exam tip

If asked for the **exact** height of the pyramid , leave your answer in surd (square root) form so your answer would be $\sqrt{82}$ m.

Exam-style questions

1 A square has a diagonal of length 10 cm. Find the length of one side of the square. Give your answer in the form $a\sqrt{b}$ cm where a and b are integers. [4]

2 Find the length of the diagonal, d, of a cube of side 8 mm. Give your answer correct to 3 significant figures. [4]

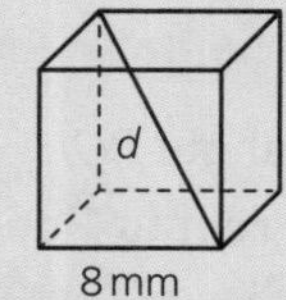

Short answers on page 139
Full worked solutions online

CHECKED ANSWERS ONLINE

Trigonometry

REVISED

Key facts

1 **Trigonometry** is about the relationship between the side lengths and the angles in a triangle.
This section is only dealing with right-angled triangles.
For any right-angled triangle:

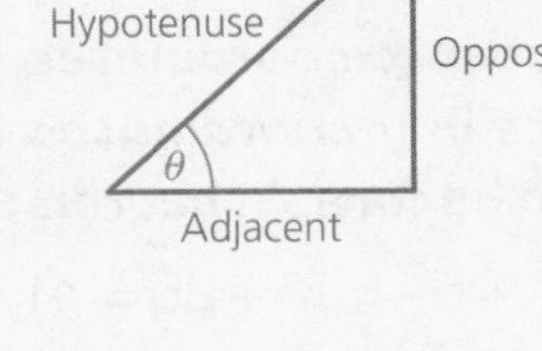

$$\sin\theta = \frac{\text{Opposite}}{\text{Hypotenuse}} = \frac{O}{H}$$

$$\cos\theta = \frac{\text{Adjacent}}{\text{Hypotenuse}} = \frac{A}{H}$$

$$\tan\theta = \frac{\text{Opposite}}{\text{Adjacent}} = \frac{O}{A}$$

2 You can use these ratios for sin, cos and tan when the triangle is **right-angled** and

- you **know two sides** and you want to find an **angle**, or
- you **know one side** and **one angle** and you want to find a **second side**.

3 You need to be able to rearrange the formulae for sin, cos and tan:

Example: $\cos\theta = \frac{A}{H}$ can be rearranged to give: $H = \frac{A}{\cos\theta}$ or $A = H \times \cos\theta$

You can use the mnemonic S OH~C AH~T OA and these cover-up triangles to help you:

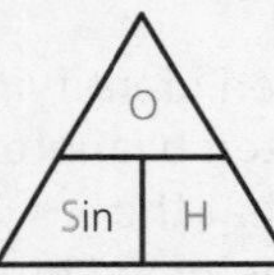

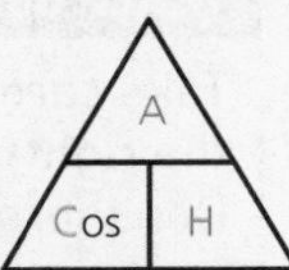

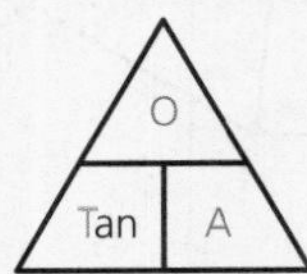

Worked examples

Finding an unknown side

Calculate the length of AB.
Give your answer correct to 3 significant figures.

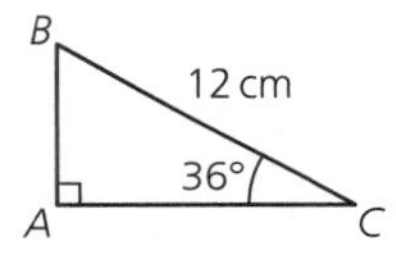

Solution

You know the hypotenuse and θ and you want the opposite side, AB so use

$$\sin\theta = \frac{O}{H}$$

B
Hypotenuse 12 cm
Opposite
36°
A
C

So $\sin 36° = \frac{AB}{12}$

Rearranging gives: $AB = 12 \times \sin 36° = 7.053\ldots$ cm

So $AB = 7.05$ cm to 3 s.f.

Finding an unknown angle

Work out the size of angle PQR.
Give your answer correct to 1 decimal place.

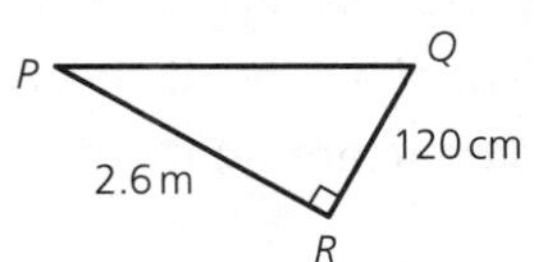

Solution

Use the same units: 120 cm = 1.2 m

You know the opposite and the adjacent sides and you want to find θ.

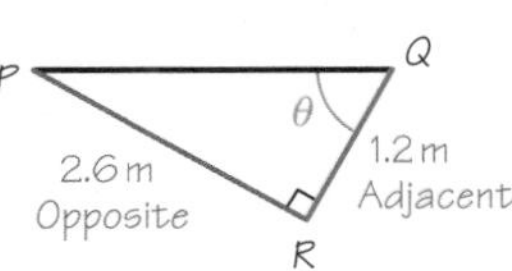

Watch out!

Make sure your calculator is in 'degrees' mode (Deg or D). Check that your calculator gives $\cos 60° = \frac{1}{2}$, if it doesn't you are in the wrong mode.

Exam tip

Label the triangle with two from Opposite, Adjacent and Hypotenuse to help you choose the right trig ratio.

Remember

The sides must have the same units.
It doesn't matter if you use cm or m as

$$\frac{2.6}{1.2} = \frac{260}{120} = 2.1\dot{6}$$

So use $\tan\theta = \frac{O}{A}$; so $\tan\theta = \frac{2.6}{1.2} = 2.1\dot{6}$

Using the $\tan^{-1}$ key: $\theta = \tan^{-1} 2.1\dot{6} = 65.22\ldots$

So angle $PQR = 65.2°$ to 1 d.p.

Using trigonometry to solve multi-step problems

The diagram shows a vertical mast, AD, supported by two straight cables, AB and AC.

Find the length of the cable AB.

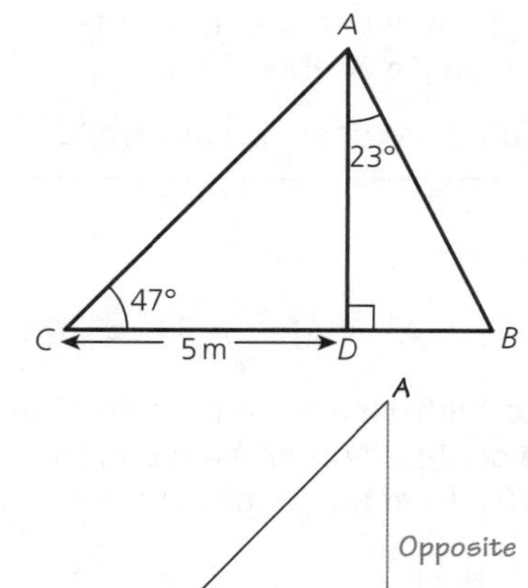

Solution

First find the height of the mast AD.

You know the adjacent and θ and you want the opposite, AD.

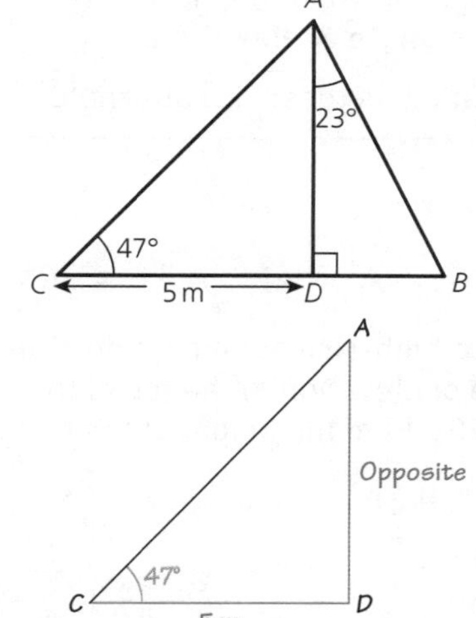

So use $\tan\theta = \frac{O}{A}$, which gives

$$\tan 47° = \frac{AD}{5}$$

Rearranging: $AD = 5 \times \tan 47° = 5.3618\ldots$ m

Now find AB: you know the adjacent and θ and you want the hypotenuse, AB.

So use $\cos\theta = \frac{A}{H}$, which gives

$$\cos 23° = \frac{5.3618\ldots}{AB}$$

Rearranging: $AB = \frac{5.3618\ldots}{\cos 23°} = 5.822\ldots$

So the cable AB is 5.82 m (to 3 s.f.)

Exam tip

You should usually give angles correct to 1 decimal place.

Watch out!

Make sure you can use the $\sin^{-1}$, $\cos^{-1}$ and $\tan^{-1}$ keys on your calculator when finding an angle. You normally have to press Shift or 2nd F key first.

Check that $\sin^{-1} 0.5 = 30$

Exam tip

Do not round in the middle of a multi-step question otherwise you may lose marks due to rounding errors. Make sure you store any answers in your calculator as you go.

Exam-style questions

1

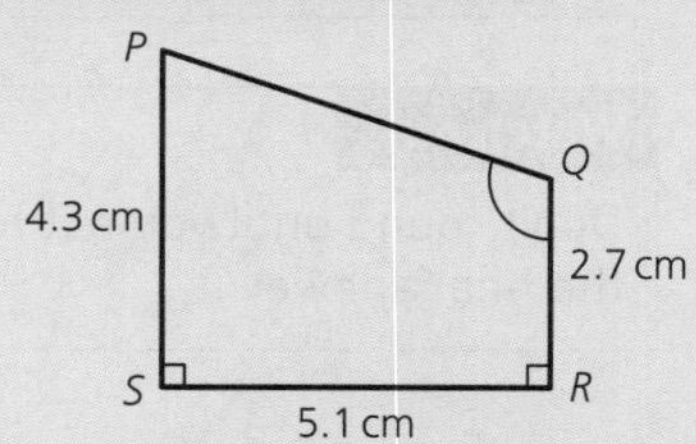

$PQRS$ is a trapezium.

Find the size of angle PQR. [4]

2

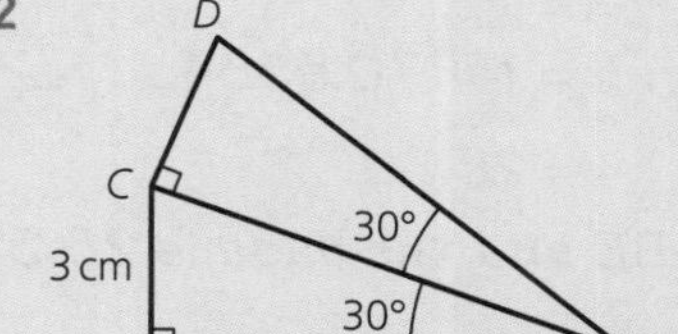

ABC and ACD are right-angled triangles.

Work out the length of AD. [5]

Short answers on page 139

Full worked solutions online

CHECKED ANSWERS ONLINE

3D trigonometry

REVISED

Key fact

You can use trigonometry to solve problems in 3 dimensions.

- If you look down at an object the angle between the horizontal and your **line of sight** is called the **angle of depression**.
- If you **look up** at an object the angle between the horizontal and your **line of sight** is called the **angle of elevation.**

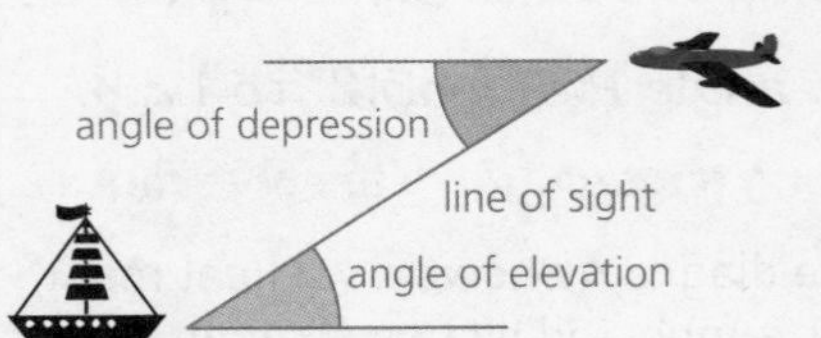

Angle of depression = angle of elevation as these are alternate angles.

Worked examples

Using angles of elevation and depression

A surveyor measures the height of a high-rise building. She stands 20 m away and measures the angle of elevation to the top of the building. The angle of elevation is 70°. Find the height of the building.

Solution

$\tan 70° = \frac{h}{20}$

So $h = 20 \times \tan 70° = 54.94\ldots$ m

Solving problems in 3D

ABCDEFGH is a cuboid.

$EH = 12$ cm, $HG = 5$ cm and $GB = 7$ cm.

Work out the angle between the diagonal *EB* and the base *EFGH*.

Solution

By Pythagoras' theorem:

$EG^2 = 12^2 + 5^2 = 169$

$EG = \sqrt{169} = 13$ cm

So $\tan\theta = \frac{7}{13} = 0.5384\ldots$

Using the $\tan^{-1}$ key: $\theta = \tan^{-1} 0.5384\ldots$

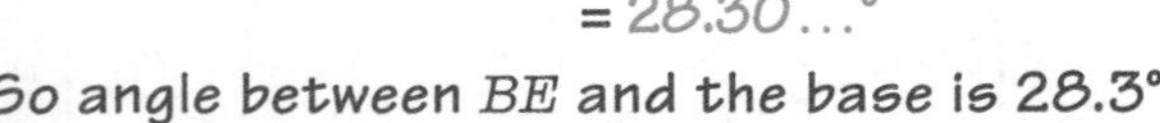

$= 28.30\ldots°$

So angle between *BE* and the base is 28.3° to 1 d.p.

Remember

Always draw a diagram.

Exam tip

Questions often combine Pythagoras' theorem and trigonometry.
Use Pythagoras' when you know two sides of a right-angled triangle and need a third side.

Watch out!

Don't round until you get to the final answer.

Exam-style question

The diagram shows a triangular prism where $AC = 12$ cm and $AD = 16$ cm.

Angle $DFE = 45°$

Angle ABC = Angle $DEF = 90°$

Calculate the size of the angle that the line *CD* makes with the plane *BEFC*.

Give your answer correct to 1 decimal place.

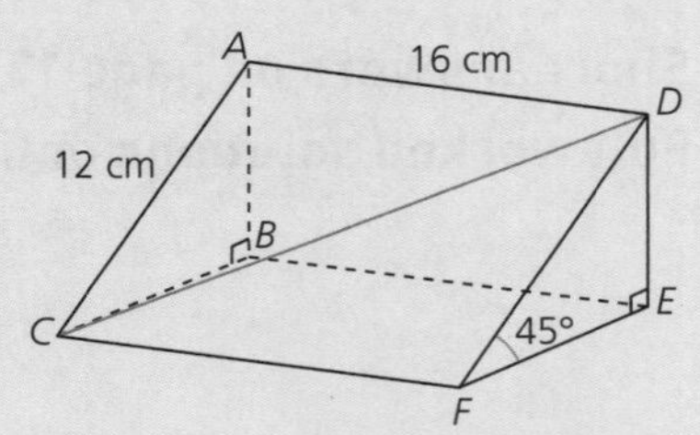

Short answers on page 139

Full worked solutions online

CHECKED ANSWERS ONLINE

The sine rule

REVISED

Key facts

1 To find missing sides or angles in triangles without right angles you need to use the **sine rule** or the **cosine rule** (see page 76).
In triangle ABC, the **side opposite** angle A is labelled a and so on.

- The **sine rule** is $\frac{a}{\sin A} = \frac{b}{\sin B} = \frac{c}{\sin C}$
  Use this form to find **missing sides.**
- You can **rewrite the sine rule as** $\frac{\sin A}{a} = \frac{\sin B}{b} = \frac{\sin C}{c}$
  Use this form to find **missing angles.**

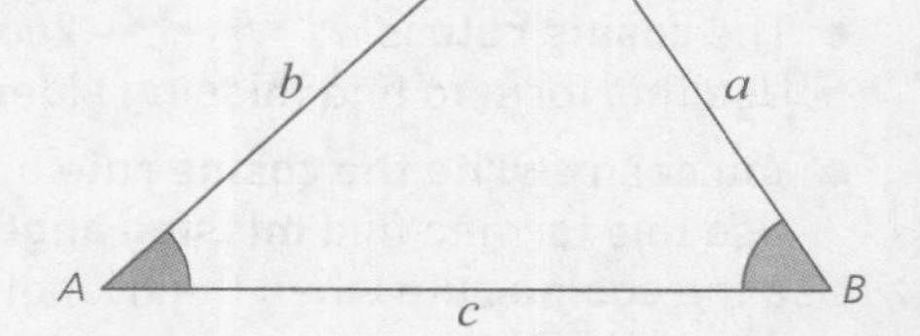

2 Use the sine rule when the triangle is not right-angled and you know

- two angles and one side and you want a second side
- two sides and an angle opposite to one of these sides and you want a second angle.

Worked examples

Using the sine rule to find a missing side

The diagram shows triangle ABC.
Calculate the length of AB.
Give your answer correct to 3 s.f.

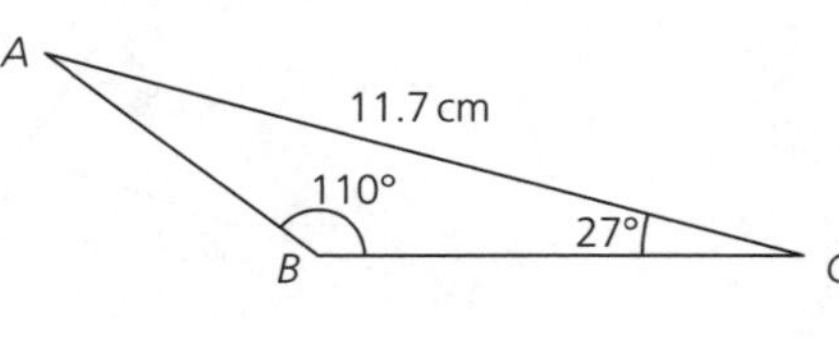

Solution

The sine rule is $\frac{c}{\sin 27°} = \frac{11.7}{\sin 110°}$

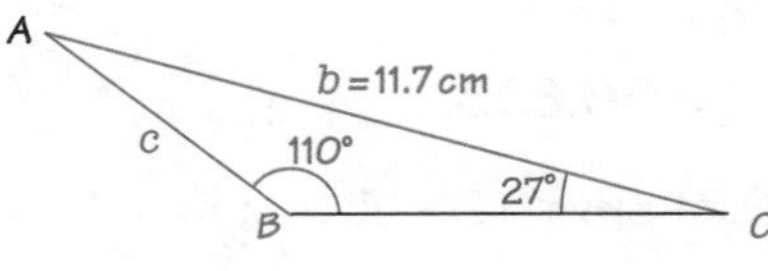

So $c = \frac{11.7}{\sin 110°} \times \sin 27° = 5.652...$

So $AB = 5.65$ cm to 3 s.f.

Using the sine rule to find a missing angle

Calculate the value of x.
Give your answer correct to 1 d.p.

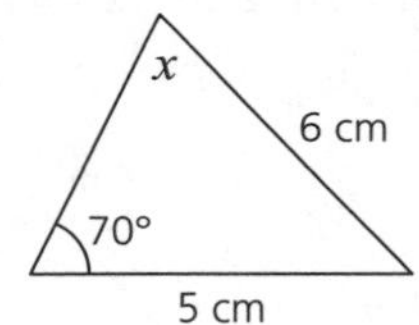

Solution

Use the sine rule for angles:

$\frac{\sin x}{5} = \frac{\sin 70°}{6}$

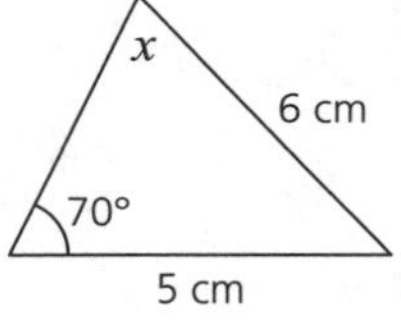

So $\sin x = \frac{\sin 70°}{6} \times 5 = 0.7830...$

So $x = \sin^{-1} 0.7830... = 51.54...° = 51.5°$ to 1 d.p.

Remember

Once you know two angles in a triangle you can work out the third angle as angles in a triangle add up to 180°
You can find BC by first working out that the angle at A is 43°

Exam tip

It is best to work this out on your calculator in one step so you avoid rounding errors.

Watch out!

If your answer for $\sin x$ is greater than 1 then you have done something wrong!

Exam-style questions

1 Calculate the value of x.
Give your answer correct 1 d.p.

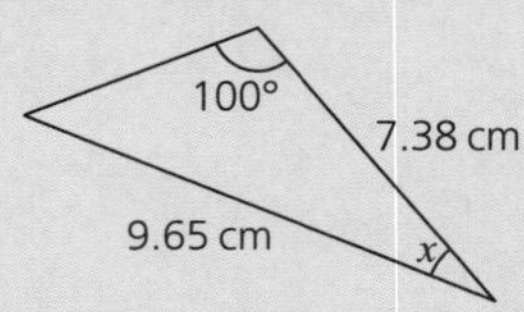

[3]

2 Calculate the length of YZ.
Give your answer correct 3 s.f.

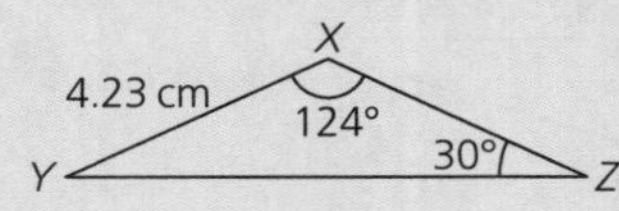

[3]

Short answers on page 139
Full worked solutions online

CHECKED ANSWERS ONLINE

The cosine rule

REVISED

Key facts

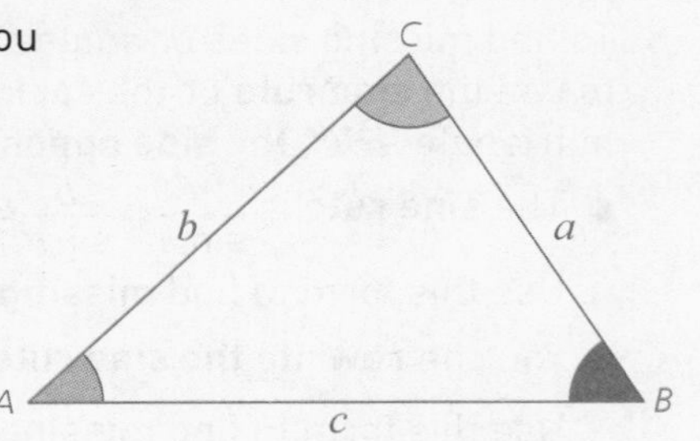

1 To find missing sides or angles in triangles without right angles you need to use the **cosine rule** or the **sine rule** (see page 75).
In triangle ABC, the **side opposite** angle A is labelled a and so on.
- The **cosine rule** is $a^2 = b^2 + c^2 - 2bc\cos A$
  Use this form to find **missing sides**.
- You can **rewrite the cosine rule** as $\cos A = \dfrac{b^2 + c^2 - a^2}{2bc}$
  Use this form to find **missing angles**.

2 Use the **cosine rule** when the triangle **is not right-angled** and you know
- two sides and angle between them and you want the third side
- three sides and you want any angle.

Worked examples

Using the cosine rule to find a missing side

The diagram shows triangle ABC.
Calculate the length of BC.
Give your answer correct to 3 s.f.

Solution

$$BC^2 = 14.2^2 + 12.7^2 - 2\times 14.2\times 12.7\cos 40°$$
$$= 362.93 - 360.68\cos 40° = 86.633\ldots$$
$$BC = \sqrt{86.633\ldots} = 9.307\ldots = 9.31 \text{ cm to 3 s.f.}$$

Using the cosine rule to find a missing angle

Triangle XYZ has sides $XY = 7$ m, $YZ = 5$ m and $XZ = 4$ m.
Find the angle XYZ.

Solution

$$\cos Y = \frac{x^2 + z^2 - y^2}{2xz}$$
$$\cos Y = \frac{5^2 + 7^2 - 4^2}{2\times 5\times 7} = \frac{58}{70} = 0.8285\ldots$$
$$Y = \cos^{-1} 0.8285\ldots = 34.04\ldots°$$

So angle $XYZ = 34.0°$ to 1 d.p.

Watch out!

A common mistake is to find $b^2 + c^2 - 2bc$ and then multiply the result by $\cos A$.

Exam tip

Make sure you show all your working and don't round until you reach your final answer.

Remember

Draw a diagram to help you.

Watch out!

If your value for $\cos A$ is not between −1 and 1 then you have made a mistake!

Exam-style questions

1 Find the size of each of the angles CAB, ABC and ACB in this triangle.
Give your answer correct to 1 d.p.

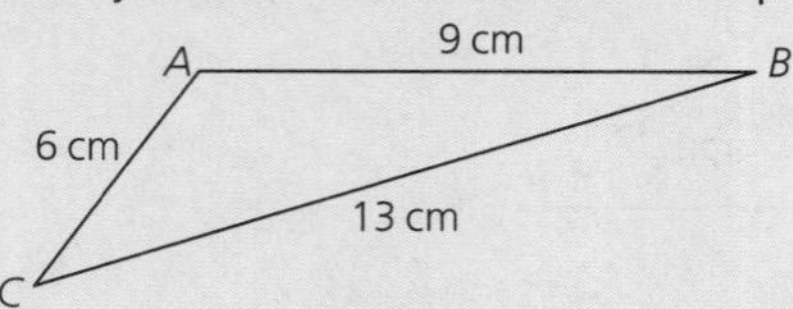

[4]

2 Calculate the value of x.
Give your answer correct to 3 s.f.

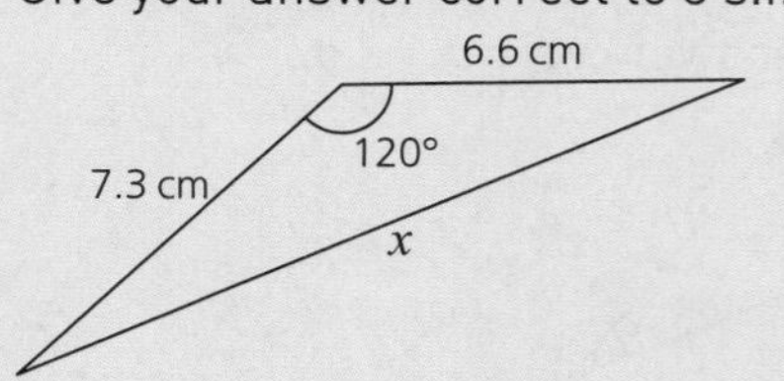

[3]

Short answers on page 139
Full worked solutions online

CHECKED ANSWERS ONLINE

The area of a triangle

REVISED

Key fact

Area of triangle $ABC = \frac{1}{2}ab\sin C$

You can use this formula to find the area of any triangle when you know two sides and the angle between them.

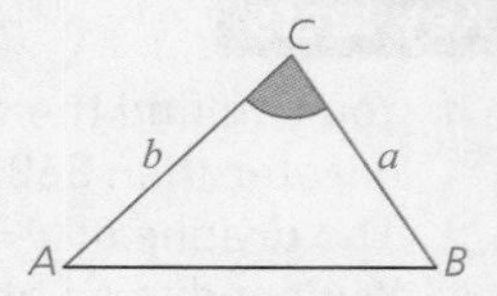

Worked examples

Finding the area of a triangle

Work out the area of the triangle.

Give your answer correct to 3 s.f.

Solution

$$\text{Area of triangle} = \frac{1}{2}ab\sin C$$
$$= \frac{1}{2}\times 6\times 9.4\sin 100^\circ = 27.77\ldots$$
$$= 27.8\text{ cm}^2 \text{ to 3 s.f.}$$

Finding the area of a segment

ABC is an arc of a circle with centre O and radius 5 cm.

AC is a chord of the circle.

Angle $AOC = 70°$

Calculate the area of the shaded segment.

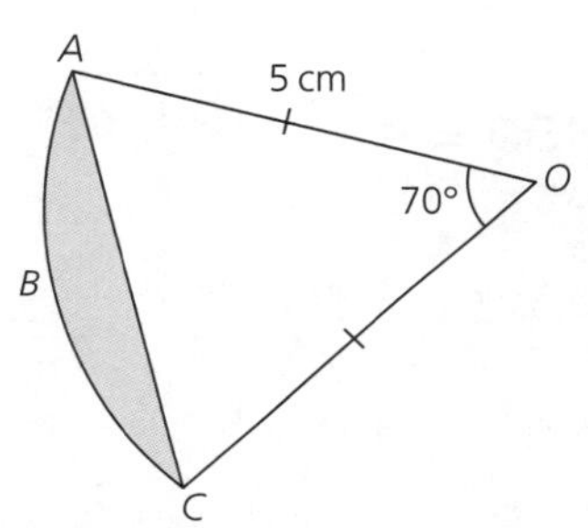

Solution

$$\text{Area of segment} = \text{area of sector } AOB - \text{area of } \triangle AOB$$
$$= \frac{70^\circ}{360^\circ}\times\pi\times 5^2 - \frac{1}{2}\times 5\times 5\sin 70^\circ$$
$$= 15.271\ldots - 11.746\ldots = 3.525\ldots$$
$$= 3.53\text{ cm}^2 \text{ to 3 s.f.}$$

Exam tip

Make sure you give your answer to the right degree of accuracy.

Remember

A chord divides a circle into two segments. See page 65 for a reminder of circles and sectors.

Watch out!

Make sure your calculator is in 'degrees' mode.
Don't forget to state the units.

Exam-style questions

1 An equilateral triangle has a side length of 10 cm.

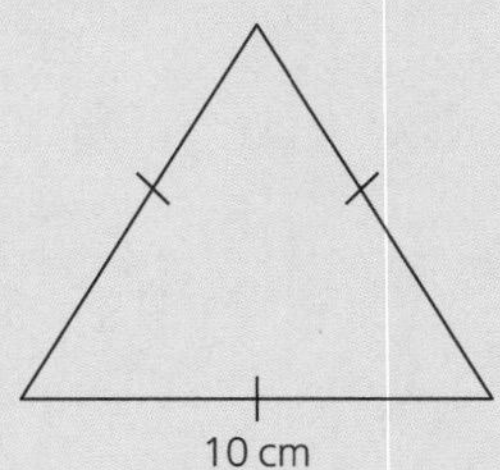

Work out the area of the triangle. [3]

2 AB is a chord of a circle with centre O.

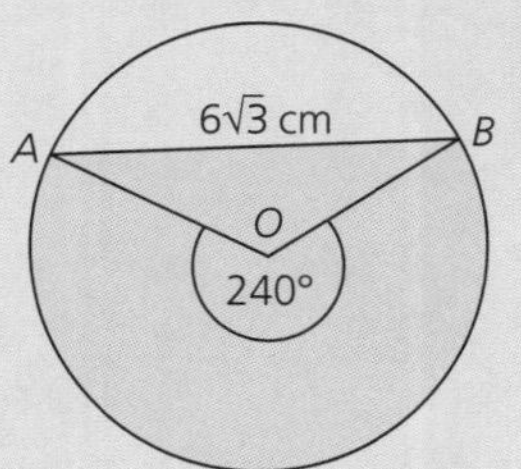

Remember

See page 76 for a reminder of the cosine rule – you'll need it for this question!

Calculate the area of the shaded region. [4]

Short answers on page 139

Full worked solutions online

CHECKED ANSWERS ONLINE

The graphs of trigonometric functions

REVISED

Key facts

1 You can find the sine, cosine and tangent of angles of any size including negative angles and angles greater than 360°
The graphs of $y = \sin x°$, $y = \cos x°$ and $y = \tan x°$ are periodic (they repeat) and have symmetry.
You need to be able recognise and draw these graphs.

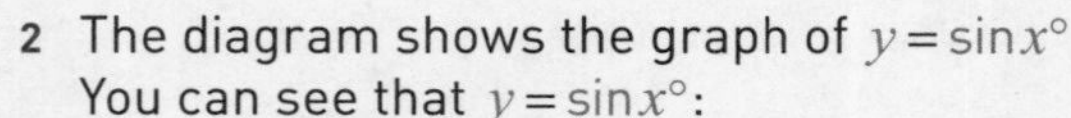

2 The diagram shows the graph of $y = \sin x°$
You can see that $y = \sin x°$:
- has rotational symmetry of order 2 about the origin
- passes the through the origin
- repeats every 360°
- lies between –1 and 1

Example: sin 90° = 1 and sin 270° = –1
- is symmetrical about $x = 90$ so $\sin x° = \sin(180 - x)°$

Example: sin 60° = sin 120°

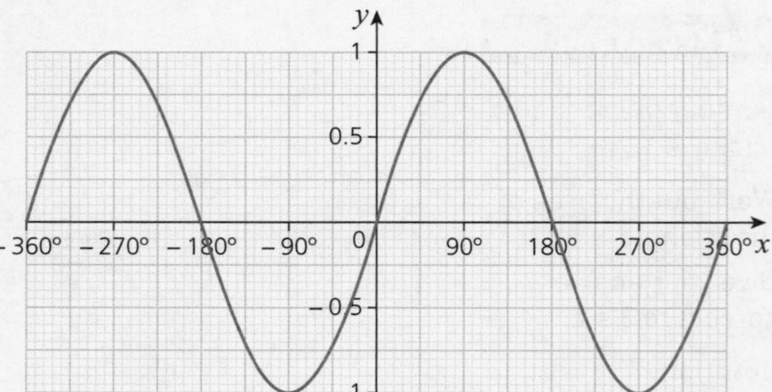

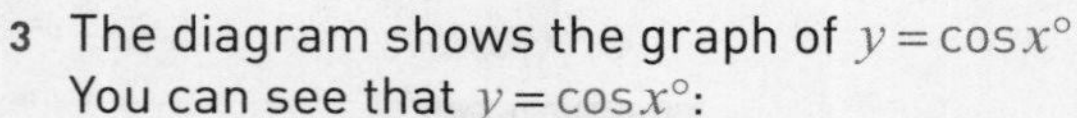

3 The diagram shows the graph of $y = \cos x°$
You can see that $y = \cos x°$:
- is symmetrical about the y-axis

Example: cos 30° = cos (–30°)
- repeats every 360°
- lies between –1 and 1

Example: cos 0° = 1 and cos 180° = –1
- is the same shape as $y = \sin x°$ (the graphs of $\sin x$ and $\cos x$ are translations of each other).

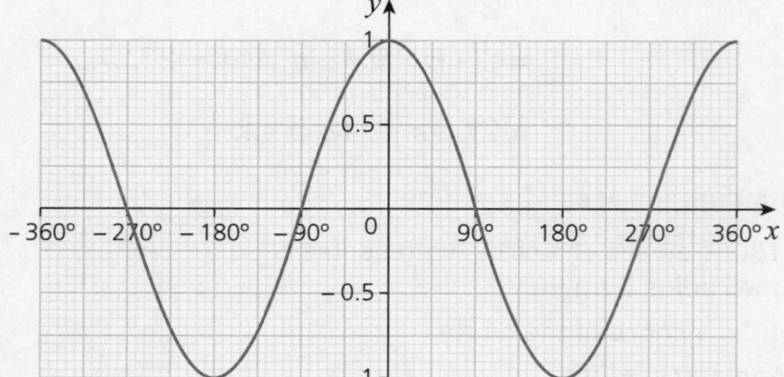

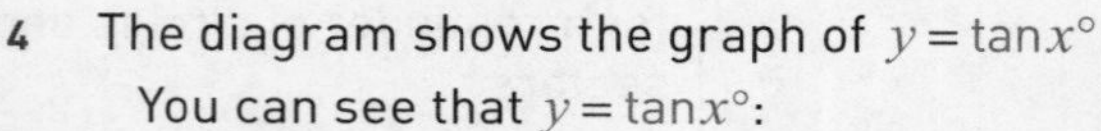

4 The diagram shows the graph of $y = \tan x°$
You can see that $y = \tan x°$:
- has rotational symmetry of order 2 about the origin
- repeats every 180°

Example: tan 30° = tan 210°
- has no maximum or minimum values
- the curve has separate 'branches' and is disconnected at $x = \pm 90, \pm 270...$ You say there are **asymptotes** (vertical lines that the curve doesn't cross) at $x = \pm 90, \pm 270...$

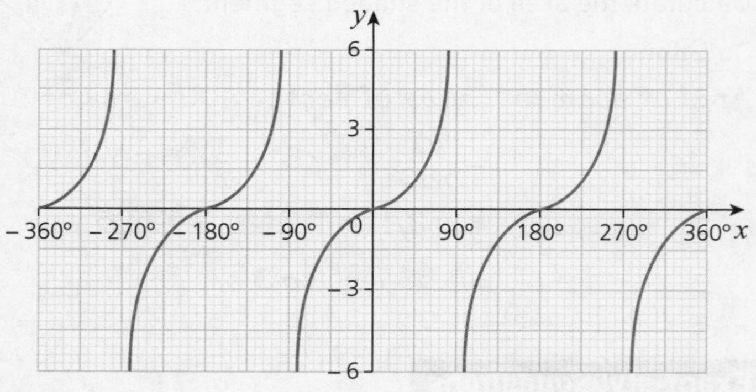

Worked examples

Using the graph of a trigonometric function

The diagram below shows the graph of $y = \cos x$

a Use the graph to estimate the value of $\cos 150°$

b Use the graph to solve i $\cos x = -0.5$ ii $4\cos x - 3 = 0$
for values of x between 0° and 360°

Solution

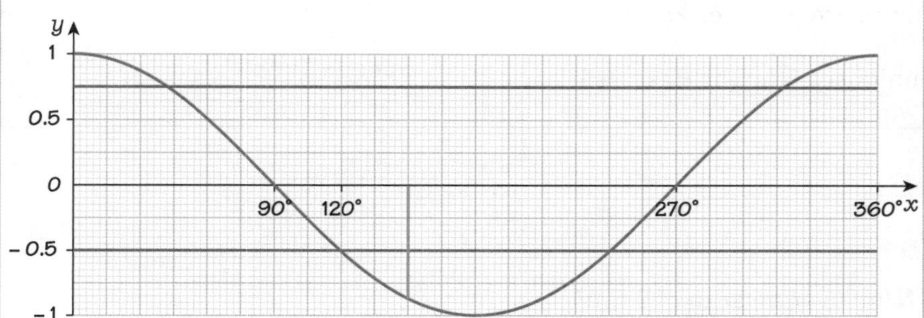

a Draw a vertical line from $x = 150$ to meet the curve.
$\cos 150° = -0.9$ to 1 d.p.

b i Draw a horizontal line from $y = -0.5$ to meet the curve twice. So $x = 150°$ or $x = 240°$

ii Rearrange $4\cos x - 3 = 0$ to give $\cos x = \frac{3}{4} = 0.75$
Draw a horizontal line from $y = 0.75$ to meet the curve twice. So $x = 40°$ or $x = 320°$

Using the fact that $\sin x° = \sin(180 - x)°$

PQR is a triangle. $PQ = 8.1$ cm and $QR = 6.3$ cm.
The area of triangle PQR is 16.4 cm^2
Find, in degrees, the two possible sizes of angle PQR.

Solution

$\frac{1}{2} \times 8.1 \times 6.3 \sin PQR = 16.4,$

so $\sin PQR = \frac{16.4 \times 2}{8.1 \times 6.3} = 0.2508...$

So $PQR = \sin^{-1} 0.2508... = 14.5°$ or $PQR = 180° - 14.5 = 165.5°$

Remember

The graph of $y = \sin x°$ makes and S shape as it passes through the origin.

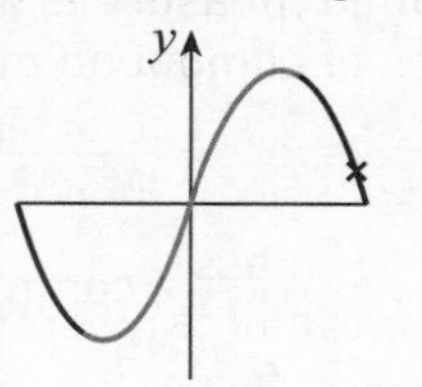

Exam tip

Add lines to the graph to help you read off the values accurately.

Watch out!

There is more than one solution to each equation – draw a line all the way across the graph to find them all!

Remember

See page 77 for a reminder of how to use $\frac{1}{2}ab\sin C$ find the area of a triangle.

Remember

$\sin x° = \sin(180 - x)°$
Always check if $180° - x$ is also a solution.

Exam-style questions

1 a Complete the table of values for $y = \tan x°$ [2]

x	0	30	45	60	75	90	105	120	135	150	180
$y = \tan x°$						–					

b Draw the graph of $y = \tan x°$ for values of x from 0° to 180° [3]

c Use your graph to solve $\tan x° = 2$ for values of x from 0° to 180° [2]

2 The sketch shows the graph of $y = \sin x°$

a Write down the coordinates of the points A, B and C. [3]

b Given $\sin 30° = \frac{1}{2}$, use the graph to find the value of

i $\sin 150°$ ii $\sin(-30)°$ [2]

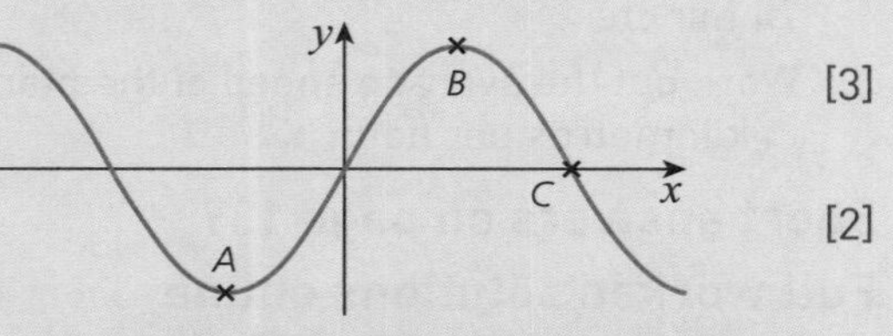

Short answers on page 139

Full worked solutions online

CHECKED ANSWERS ONLINE

Compound measures

REVISED

Key facts

1. A compound measure is made up from two or more measures.
2. Examples of compound measures are:
   - Average speed = $\frac{\text{total distance}}{\text{total time}}$, common units are km/h or m/s
   - Density = $\frac{\text{mass}}{\text{volume}}$, common units are g/cm^3 or kg/m^3
   - Pressure = $\frac{\text{force}}{\text{area}}$, common units are N/m^2 or Pascals (Pa)

Remember

DISTance
Is
Speed × Time

Worked examples

Solving problems involving pressure

A force exerts a pressure of 2500 N/m^2 over an area of 0.5 m^2.

Find the pressure exerted by the same force when it is applied over an area of 0.2 m^2.

Solution

Use Pressure = $\frac{\text{force}}{\text{area}}$ to find the force.

$2500 = \frac{\text{force}}{0.5}$, so force = $2500 \times 0.5 = 1250$ N

Now find the new pressure: Pressure = $\frac{1250}{0.2} = 6250$ N/m^2

Solving problems involving density

An alloy is made from mixing together two metals.

Metal X: density = 2500 kg/m^3; Metal Y: density = 8800 g/cm^3.

300 kg of Metal X and 550 kg of Metal Y are mixed to make the alloy.

Work out the density of the alloy. Give your answer correct to 3 s.f.

Solution

Use Volume = $\frac{\text{mass}}{\text{density}}$ to find the volume of each metal:

X: $V = \frac{300}{2500} = 0.12\,\text{m}^3$; Y: $V = \frac{550}{8800} = 0.0625\,\text{m}^3$

Total Volume = $0.12\,\text{m}^3 + 0.0625\,\text{m}^3 = 0.1825\,\text{m}^3$

Total mass = 550 kg + 330 kg = 880 kg

Total density = $\frac{880}{0.1825} = 4821...\,\text{kg/m}^3 = 4820\,\text{kg/m}^3$ to 3 s.f.

Exam tip

You will be given the formula for pressure if you need it in the exam.

Watch out!

Take care rearranging the formula.

You may find a cover-up triangle helps:

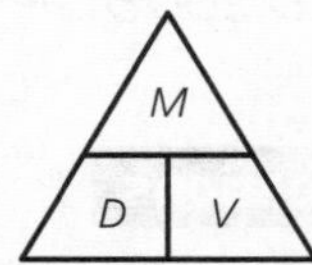

Exam-style questions

1. A plane flew from Paris to New York. The distance the plane flew was 5830 km.

   The time taken by the plane was 8 hours 24 minutes.

   Work out the average speed of the plane, in kilometres per hour.

   Give your answer correct to 3 significant figures. [3]

2. The radius of a metal sphere is 4 cm. The density of the sphere is 2.56 g/cm^3.

   Calculate the mass of the sphere. Give your answer correct to 3 significant figures. [3]

Short answers on page 139

Full worked solutions online

CHECKED ANSWERS ONLINE

Review questions: Shape, space and measure

1 The diagram shows a triangle and a rectangle. All lengths are in centimetres.

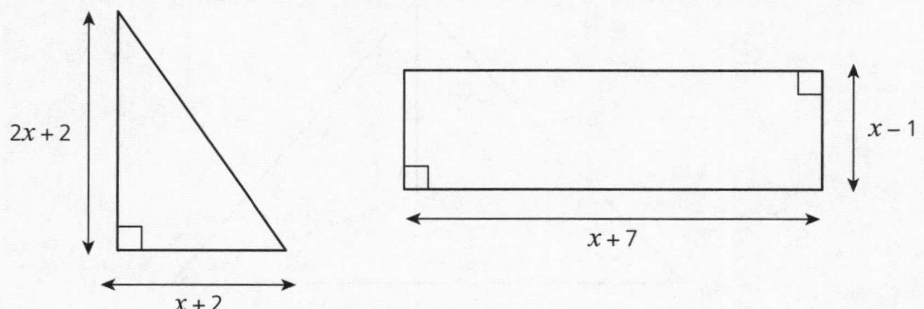

The area of the triangle is the same as the area of the rectangle.
Work out the perimeter of the rectangle. (4 marks)

2 The diagram shows a circle.
AD is the diameter of the circle with centre B.
C is the midpoint of BD.

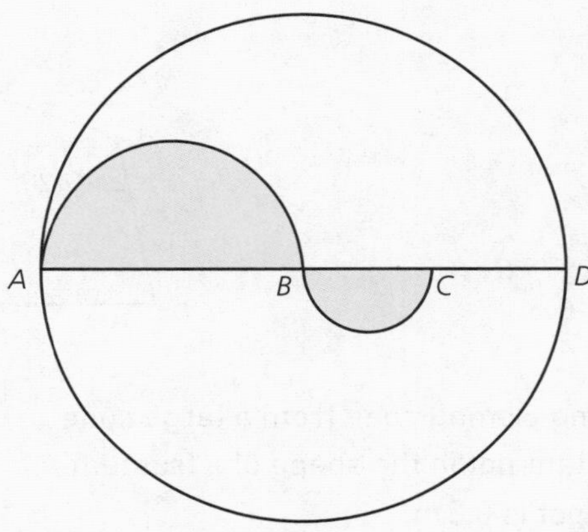

What fraction of the circle is shaded? (5 marks)

3 The diagram shows triangle ABC.

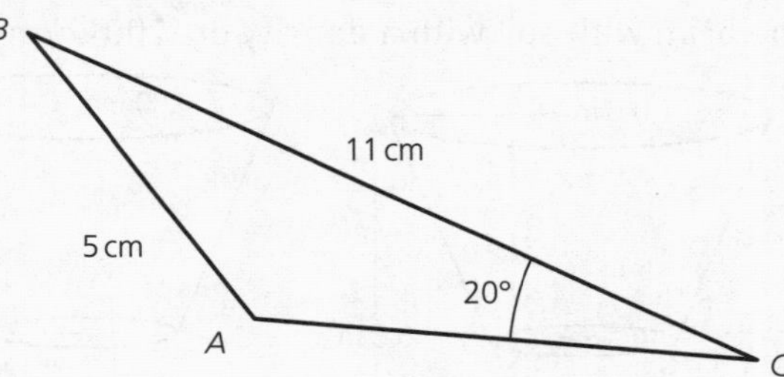

$AB = 5$ cm and $BC = 11$ cm.
Angle $ACB = 20°$
Work out the size of the obtuse angle BAC. (3 marks)

4 A flagpole, AT, is tethered by two cables TB and TC.
The points A, B and C are on horizontal level ground.

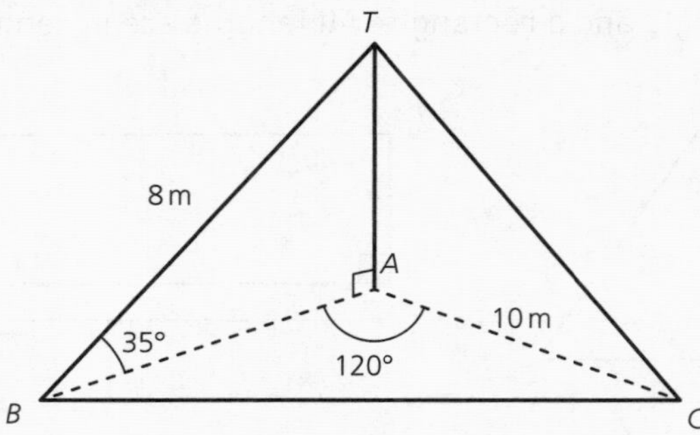

$BT = 8$ m
$AC = 10$ m
Angle $ABT = 35°$
Angle $BAC = 120°$

a Find the height of the flagpole. (3 marks)
b Find the distance BC. (4 marks)

5 $ABCD$ is a quadrilateral.
$AB = 10.3$ cm
$BC = 8.3$ cm
$CD = 11.7$ cm
Angle $BCD = 80°$
Angle $DAB = 120°$
Calculate the area of $ABCD$. (5 marks)

6 A frustum is made by removing a small cone from a large cone.
The diagram shows a large plant pot in the shape of a frustum.
The radius of the base of the pot is 0.3 m.
The radius of the top of the pot is 0.6 m.
The height of the original cone is 0.8 m.
The plant pot is filled to the brim with soil with a density of 1200 kg/m^3.

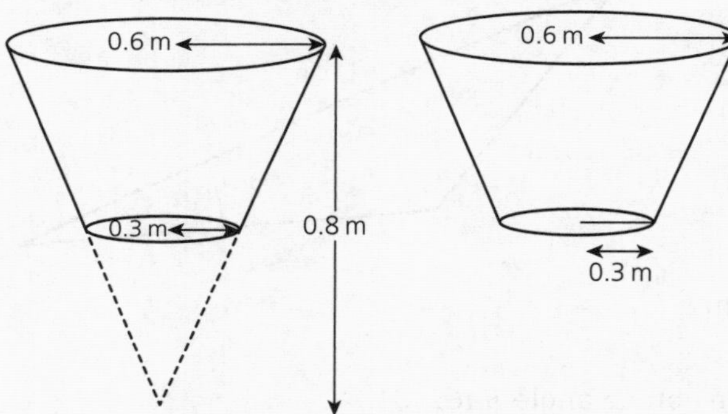

a Find the mass of soil in the pot. (4 marks)
The outside of the plant pot is painted.
b Find the total area of the plant pot that is painted.
Give your answer in the form $\frac{a}{b}\pi$ m^2 where $\frac{a}{b}$ is a fraction in its simplest form. (4 marks)

Short answers on page 139
Full worked solutions online

CHECKED ANSWERS ONLINE

Target your revision: Geometry and transformations

Check how well you know each topic by answering these questions. If you struggle, go to the page number in brackets to revise that topic.

1 Work with angle properties

XYZ and VYW are straight lines. $XY = VY$.

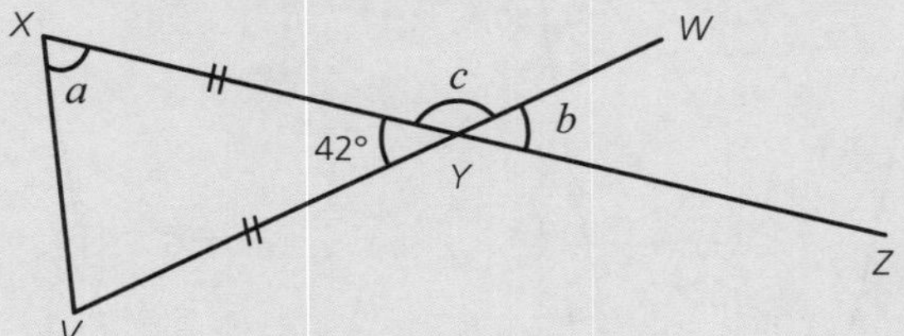

a Work out the size of angle a. Give reasons for your answer.

b Write down the size of angle b. Give a reason for your answer.

c Work out the size of angle c. Give a reason for your answer.

(see page 86)

2 Calculate angles in triangles and quadrilaterals

$ABCD$ is a parallelogram. $BCFE$ is a rhombus.

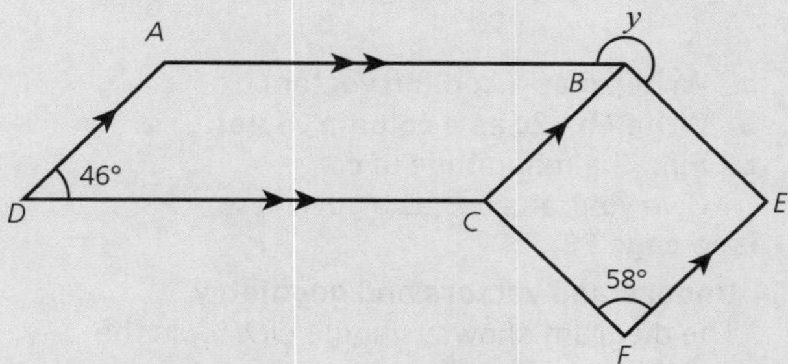

Calculate the size of the angle marked y. Give reasons for your answer.

(see page 86)

3 Understand bearings

The diagram shows the position of two towns A and B.

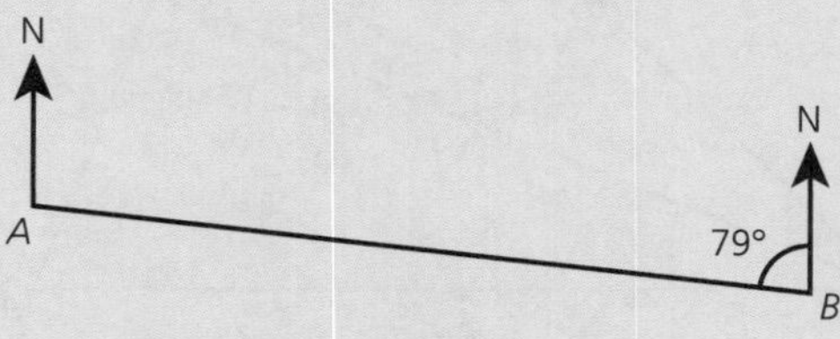

Work out the bearing of B from A.

(see page 88)

4 Understand symmetry of 2D shapes

For the shape shown, write down:

a The number of lines of symmetry.

b The order of rotational symmetry.

(see page 85)

5 Calculate angles in a polygon

Here is shape $ABCDE$. All angles are in degrees.

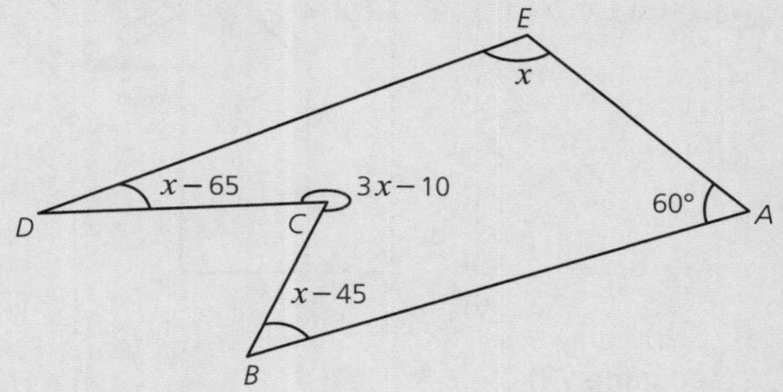

Work out the value of x.

(see page 89)

6 Understand circle theorems

A and B are points on a circle centre O. AC is a tangent to the circle and OBC is a straight line. Angle $ACO = 36°$

Work out the size of angle ABC. Give reasons for your answer.

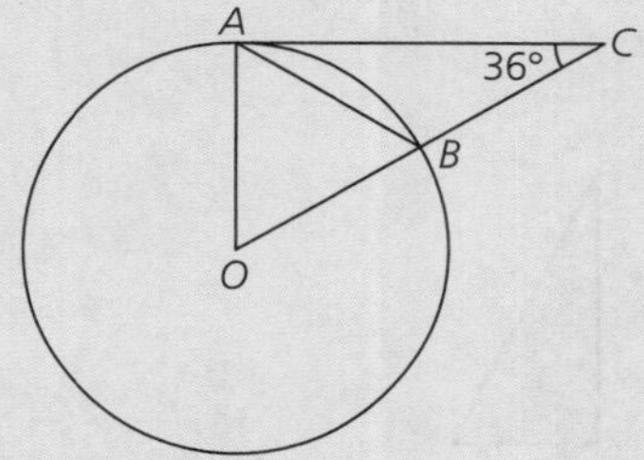

(see page 102)

7 Carry out constructions

Using ruler and compasses, construct the bisector of angle ABC.

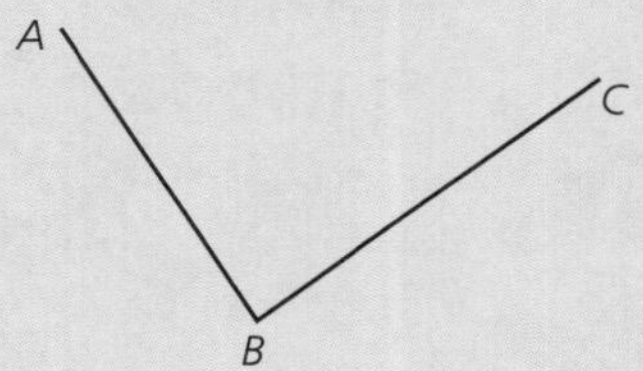

(see page 90)

8 Translate a shape

Describe fully the single transformation that maps shape **P** onto shape **Q**.

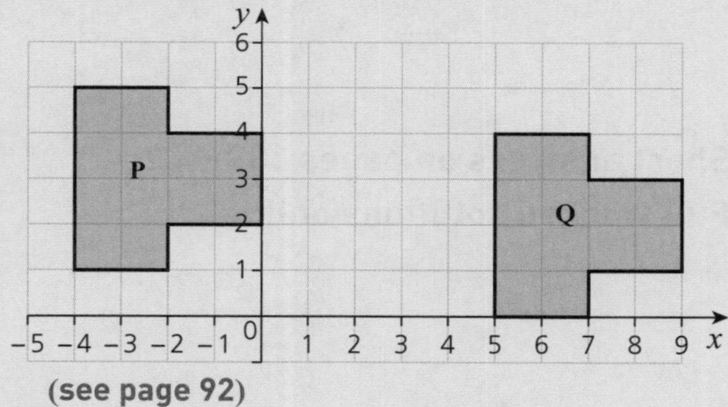

(see page 92)

9 Reflect a shape

Reflect shape **Q** in the line $x = 2$.

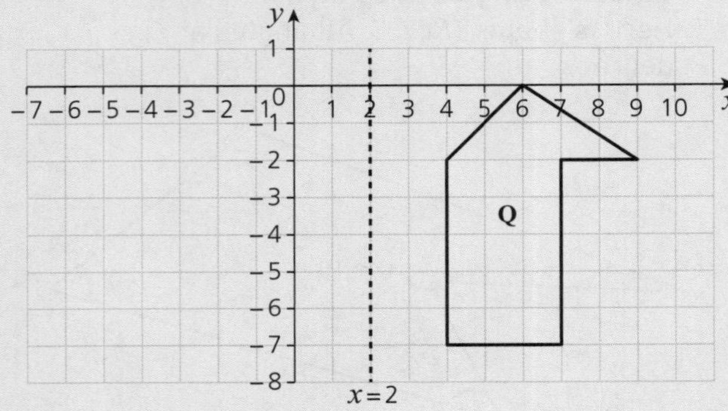

(see page 93)

10 Rotate a shape

A quadrilateral **Q** has vertices at (2, −1), (8, −1), (10, −6) and (2, −6).

Rotate the quadrilateral **Q** 270° anticlockwise about (2, 0).

(see page 94)

11 Describe an enlargement

Describe fully the single transformation that maps triangle **P** onto triangle **Q**.

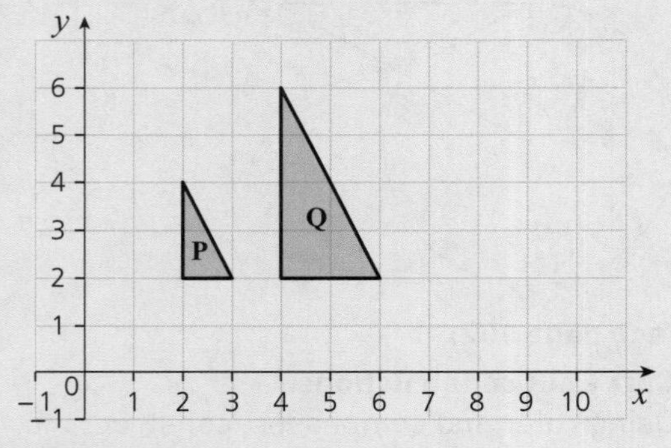

(see page 95)

12 Understand transformation of graphs

The graph of $y=\mathrm{f}(x)$ is shown. The graph **G** is a translation of $y=\mathrm{f}(x)$. Write down the equation of graph **G**.

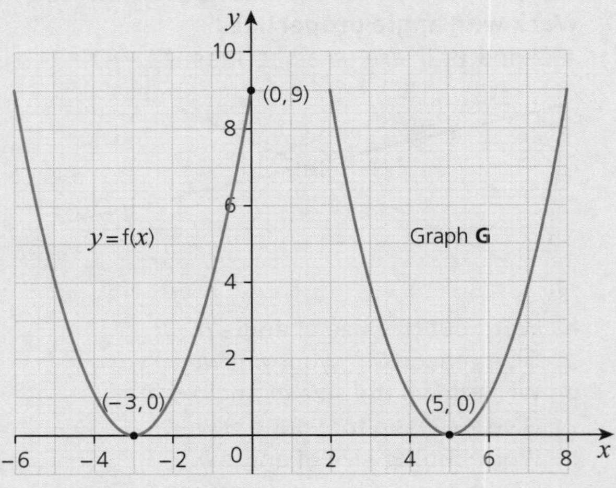

(see page 96)

13 Perform calculations with vectors

$$\mathbf{a}=\begin{pmatrix}3\\-1\end{pmatrix}\quad \mathbf{b}=\begin{pmatrix}1\\9\end{pmatrix}\quad \mathbf{c}=\begin{pmatrix}-2\\-5\end{pmatrix}$$

a Write $2\mathbf{a}$ as a column vector.

b Write $4\mathbf{b}-2\mathbf{c}$ as a column vector.

c Find the magnitude of **c**
Give your answer as a surd.

(see page 98)

14 Understand vectors and geometry

The diagram shows triangle OQP. X is the midpoint of $\overrightarrow{QP}$. $\overrightarrow{OQ}=3\mathbf{b}$ and $\overrightarrow{OP}=4\mathbf{a}$

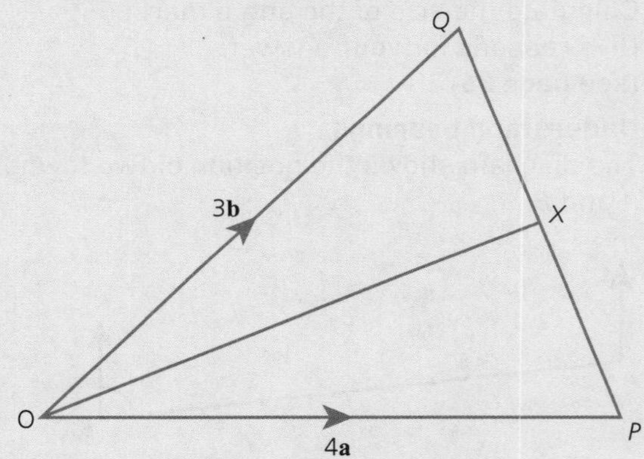

a Work out $\overrightarrow{OX}$.

Y is on $\overrightarrow{OX}$ such that $OY:YX=2:1$

b Work out $\overrightarrow{PY}$.

(see page 100)

Short answers on pages 139–140
Full worked solutions online

CHECKED ANSWERS ONLINE

2D shapes

REVISED

Key facts

1 **Order of rotational symmetry** is the number of times a shape looks the same when rotated a complete turn.

2 **Square:**
- 4 lines of symmetry
- Rotational symmetry order 4

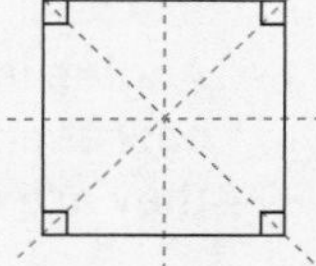

3 **Rectangle:**
- 2 lines of symmetry
- Rotational symmetry order 2

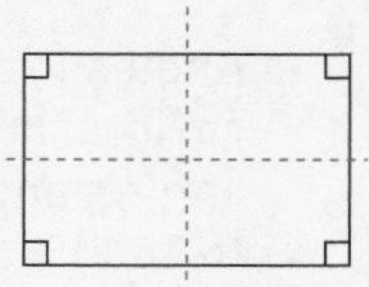

4 **Rhombus:**
- Two lines of symmetry
- Rotational symmetry order 2
- 4 equal sides
- 2 pairs of equal angles
- 2 pairs of parallel sides
- Diagonals bisect at right angles

5 **Parallelogram**
- No lines of symmetry
- Rotational symmetry order 2
- 2 pairs of equal sides
- 2 pairs of equal angles
- 2 pairs of parallel sides
- Diagonals bisect each other

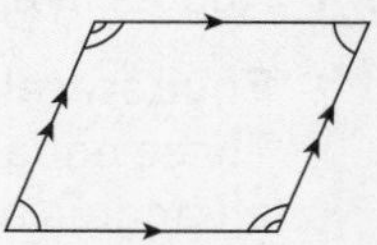

6 **Trapezium:**
- No (or 1) lines of symmetry
- No rotational symmetry
- 1 pair of parallel sides

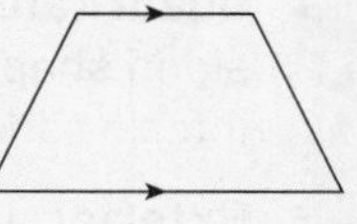

7 **Kite:**
- 1 line of symmetry
- No rotational symmetry
- 2 pairs of equal sides
- 1 pair of equal angles
- Diagonals cross at right angles

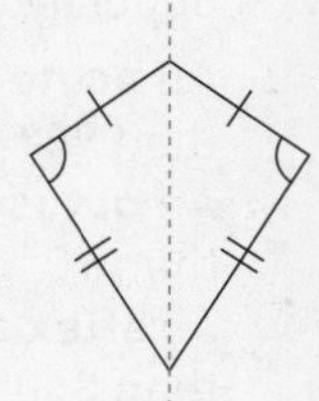

Worked examples

Finding rotational symmetry

This shape has rotational symmetry.

Write down the order of rotational symmetry.

Solution

When the shape is rotated a complete turn it looks exactly the same 3 times. So the order of rotational symmetry is 3.

Giving a shape rotational symmetry

Add three squares to make a shape with rotational symmetry of order two and two lines of symmetry.

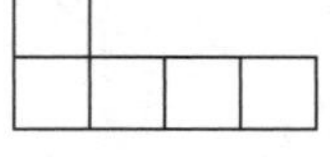

Solution

Three squares can be added as shown. The shape looks exactly the same twice when rotated a complete turn.

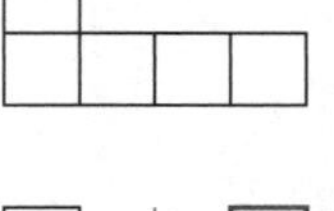

Exam tip

Draw a circle on a corner of the shape. Trace the shape and rotate about its centre to determine the order of rotation.

Watch out!

When a shape fits onto itself only once in a full rotation then it doesn't have rotational symmetry and you say it has 'order 1'.

Remember

The order of rotational symmetry of a regular polygon is equal to the number of sides.

Exam-style question

Here are some shapes.

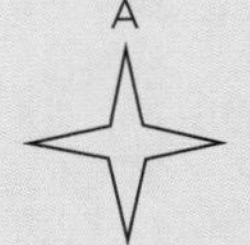

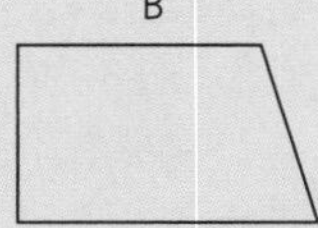

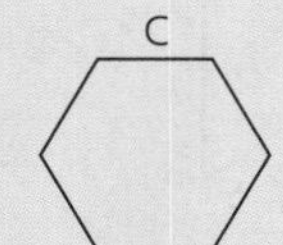

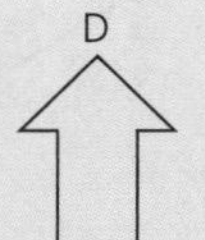

a Which shape has no line of symmetry? [1]

b Which shape has one line of symmetry? [1]

c Write down the shape that has

i rotational symmetry of order 4

ii rotational symmetry of order 6 [2]

Short answers on page 140

Full worked solutions online

CHECKED ANSWERS ONLINE

Angles

REVISED

Key facts

1 Angles in a triangle add up to 180°

2 **Equilateral triangle**: Three equal sides and all angles are 60°

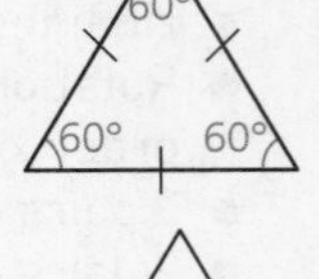

3 **Isosceles triangle**: 2 sides are equal and 2 base angles are equal.

4 **Quadrilateral**: A four-sided shape with sum of interior angles = 360°

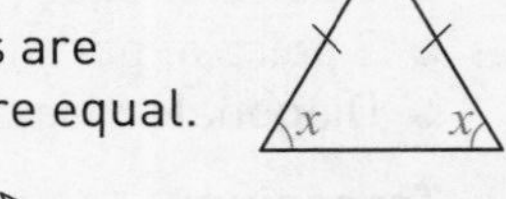

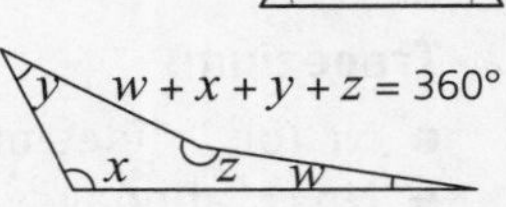

5 **Exterior angle** is equal to the sum of the interior opposite angles. So $z=x+y$

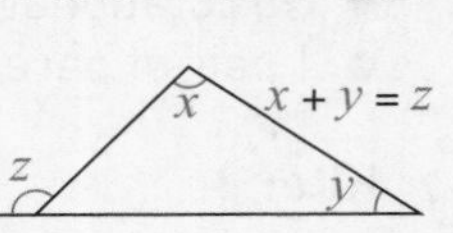

6 An **acute angle** is less than 90°
An **obtuse angle** is greater than 90° but less than 180°
A **reflex angle** is greater than 180° but less than 360°

7 **Parallel lines** are indicated by lines with arrows:

8 Angles in a Z shape are **alternate angles** and they are **equal**.

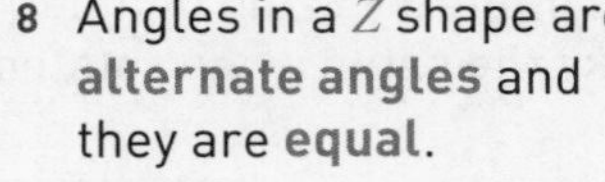

9 Angles in an F shape are **corresponding angles** and they are **equal**.

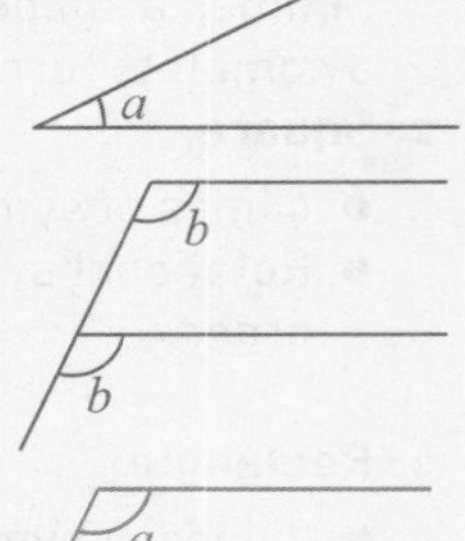

10 Angles in a C shape are **co-interior** (or **allied**) **angles** and they **add up to 180°**

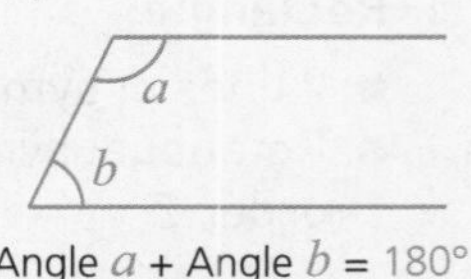

Angle a + Angle b = 180°

11 **Vertically opposite angles** are **equal**.

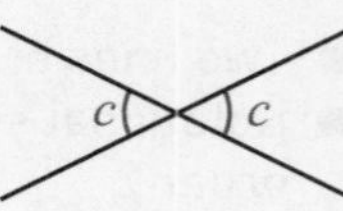

12 **Angles on a straight line** add up to **180°**

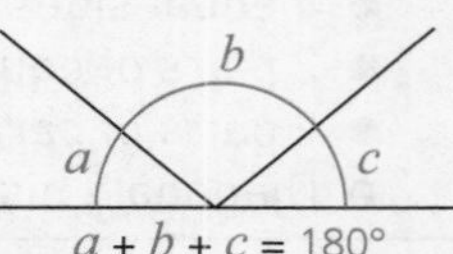

13 **Angles around a point add up to 360°**

Worked examples

Missing angles and algebra

The diagram shows quadrilateral $ABCD$. Work out the size of the obtuse angle in the quadrilateral.

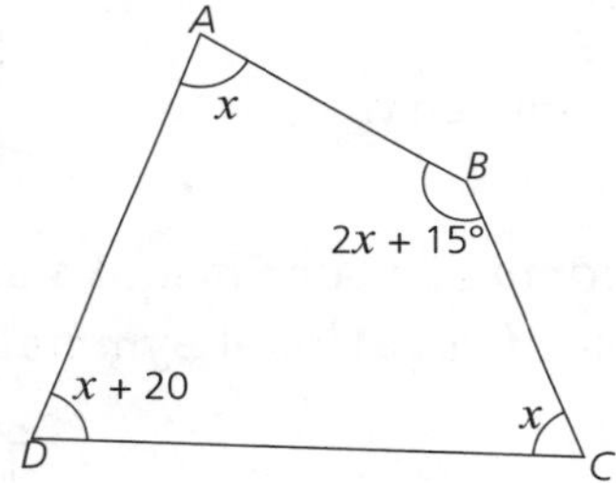

Solution

$$x + 2x + 15 + x + x + 20 = 360$$
$$5x + 35 = 360$$
$$5x = 325$$
$$x = 65$$

The obtuse angle is at B, so $2 \times 65 + 15 = 145°$

Finding angles on a straight line

ABC and DBE are straight lines.

a Write down the size of angle m.

b i Work out the size of angle n.

ii Give a reason for your answer.

c $BC=BE$. Work out the size of angle p.

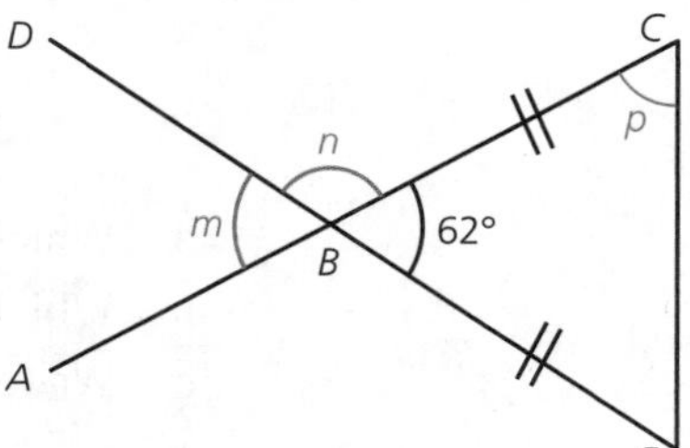

Remember

The sum of the interior angles of a quadrilateral is 360°

Watch out!

Do not measure angles as they won't be drawn accurately.

Exam tip

Make sure you answer the whole question, you need to work out the size of the obtuse angle, not just find the value of x

Solution

a $m = 62$ as vertically opposite angles are equal.

b i $62° + n = 180° \Rightarrow n = 180° - 62° = 118°$

ii angles on straight line add to 180°

c $BC = BE$ which means BCE is an isosceles triangle so $p = (180 - 62) \div 2 = 59°$

Finding angles on parallel lines

The lines AB and CD are parallel.

Find the angles:

i s ii t iii u

Give reasons for your answer in each case.

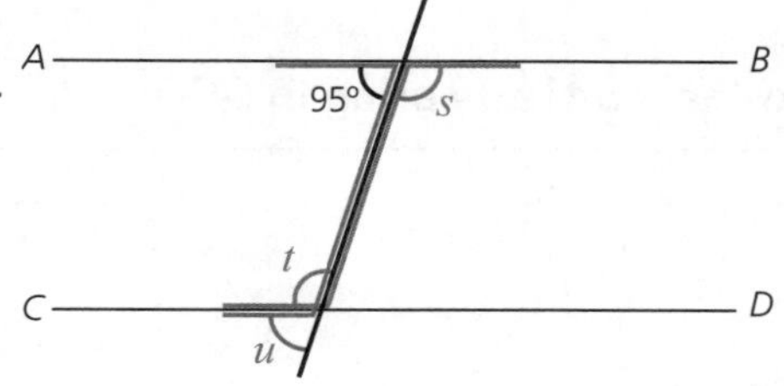

Solution

i $s = 180° - 95° = 85°$ as angles on straight line add up to 180°

ii $t = s = 85°$ as alternate angles are equal

OR $t = 85°$ as co-interior angles add to 180°

iii $u = 180° - 85° = 95°$, as angles on a straight line add up to 180°

OR $u = 95°$ as corresponding angles are equal

> **Watch out!**
>
> **Correct names** for angle types must be given i.e. **alternate, corresponding**, etc. not **Z** shape or **F** shape etc.

> **Exam tip**
>
> When a reason is asked for an actual written reason is required, not just working out.

> **Remember**
>
> There can be more than one method when finding angles involving parallel lines.

Exam-style questions

1 In the diagram AE is parallel to BD.

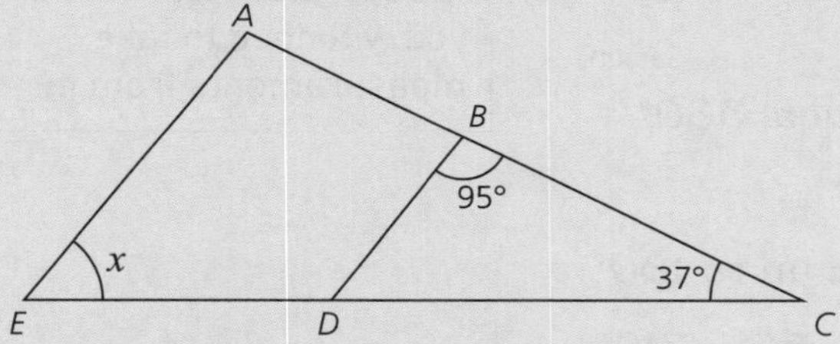

Work out the size of the angle marked x. [2]

2 The lines ABC and DE are parallel.

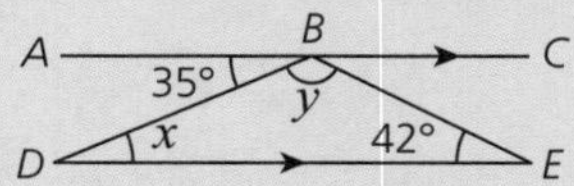

Work out the value of x and the value of y.

Give reasons for your answers. [4]

3 ABC is a triangle. BCD is a straight line.

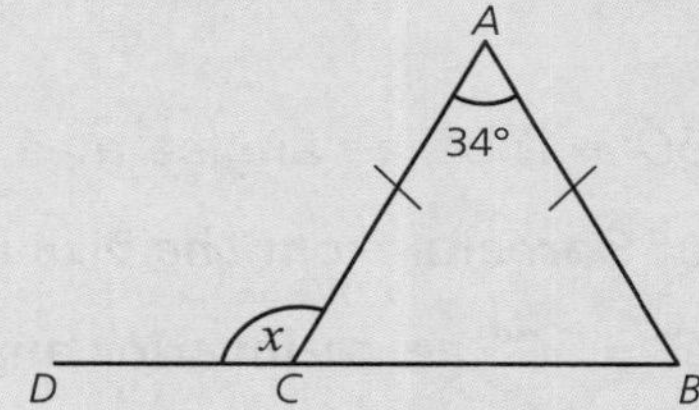

Work out the value of x. [2]

4 $ABCD$ is a quadrilateral.

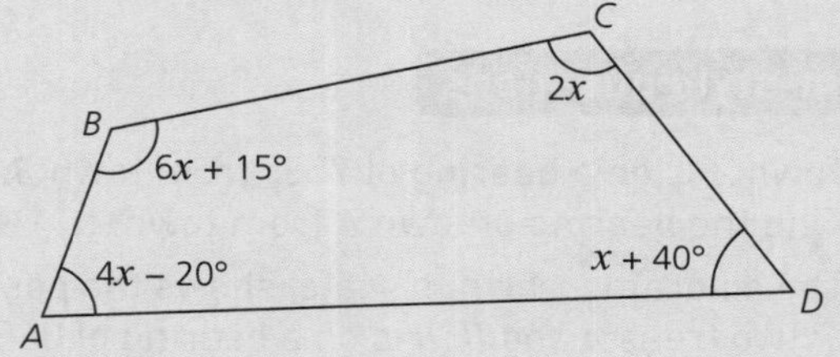

a Find the value of x. [3]

b Which angle is the smallest? [2]

Short answers on page 140

Full worked solutions online

CHECKED ANSWERS ONLINE

Bearings

REVISED

Key facts

Bearings are measured clockwise from the North line.
Always use 3 figures to write a bearing.
Add a zero in front of bearings less than 100°

To measure the bearing of P from A, place a protractor with 0° on the North line at A.
Measure the angle going clockwise, so the bearing is 060°

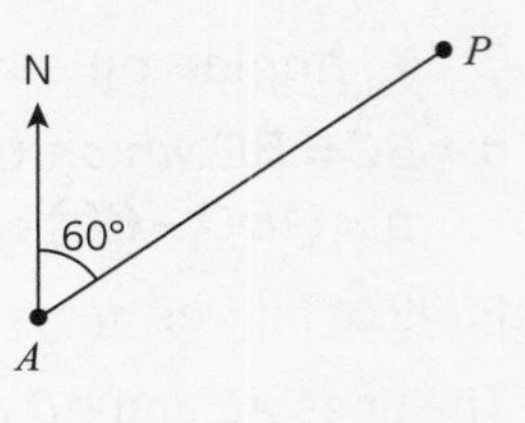

Worked examples

Finding a bearing

Saloni walks from point A on a bearing of 047°. She walks for two miles in a straight line to point B. Find the bearing of A from B.

Solution

First, sketch a diagram.
$x = 180° - 47° = 133°$ as co-interior angles sum to 180°, so the bearing of A from $B = 360° - 133° = 227°$

Finding multiple bearings

The diagram (not drawn to scale) shows the position of Sareena and her bike.

a Work out the bearing of Sareena from her bike.
b Work out the bearing of the bike from Sareena.

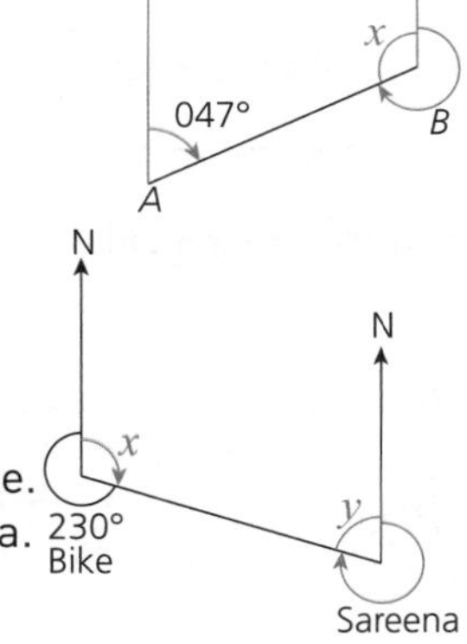

Solution

a $x = 360° - 230° = 130°$, *as angles at a point equal 360°*

So the bearing of Sareena from the bike is 130°

b $y = 180° - 130° = 50°$ *as co-interior angles add up to 180°*

So the bearing of the bike from Sareena is $360° - 50° = 310°$ as angles at a point add up to 360°

Remember

Always start North and go clockwise.

Watch out!

Make sure you work out the correct angle!

Exam tip

Angles can be measured **only** if the diagram is drawn to scale. If the exam questions say 'Diagram is accurately drawn' then you will need to take measurements from it.

Exam-style questions

1 Town A is on a bearing of 255° from town B. Find the bearing of town B from town A. [2]

2 The diagram, drawn to scale, shows the position of two trees A and B. B is on a bearing of 065° from A. A third tree C is on a bearing of 115° from A and 215° from B. Mark the position of the third tree C. [2]

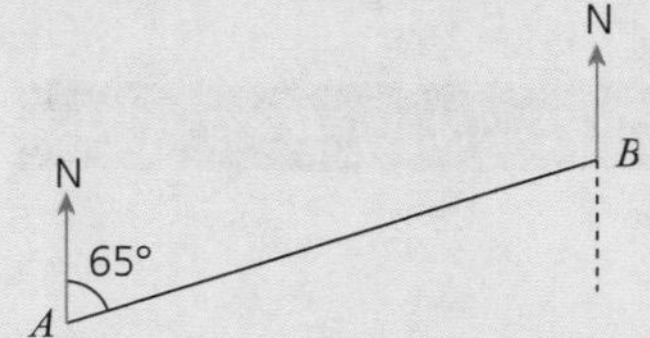

3 The diagram shows the position of three houses A, B and C. The bearing of B from A is 128°. C is due West of B. $AB = BC$. Calculate the bearing of A from C. [3]

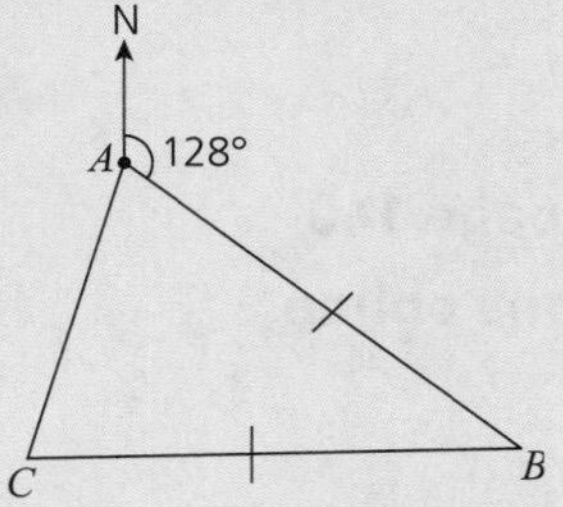

Short answers on page 140
Full worked solutions online

CHECKED ANSWERS ONLINE

Angles in polygons

REVISED

Key facts

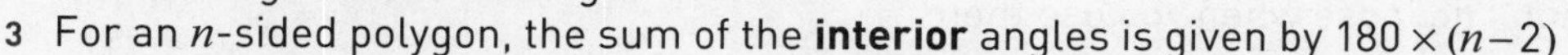

1. A **polygon** is a 2D shape with straight sides. A **regular polygon** has all sides the same length and all angles equal.
2. A polygon has an **interior angle** (inside) and **exterior angle** (outside).
   interior angle + exterior angle = 180°
3. For an n-sided polygon, the sum of the **interior** angles is given by $180 \times (n-2)$
4. The sum of **exterior** angles of an n-sided polygon = 360°
5. The **exterior angle**, E, of an n-sided polygon is given by $E = \frac{360}{n}$

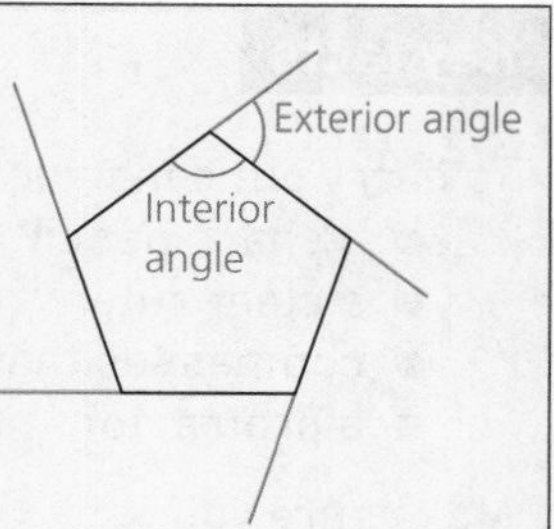

Worked examples

Finding the number of sides of a polygon

Each interior angle of a regular polygon is 144°

Work out the number of sides of the polygon.

Solution

Exterior angle $x = 180° - 144° = 36°$

Number of sides $= \frac{360°}{36°} = 10$

Finding a missing angle using algebra

Here is a hexagon. Work out the value of x.

Solution

Sum of interior angles $= 180° \times (6-2) = 720°$

$127 + 111 + 153 + x + 135 + x = 720$

$2x + 526 = 720$

$2x = 194$

$x = 97°$

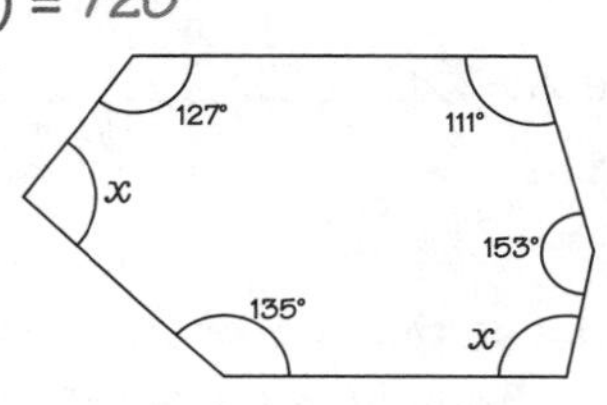

Solving problems

The interior angle of a regular polygon is 14 times larger than its exterior angle.

Work out the sum of the interior angles of the polygon.

Solution

$14x + x = 180°$, so $15x = 180°$

and $x = 12°$

Number of sides $= 360° \div 12° = 30$

Sum of interior angles $= (30 - 2) \times 180° = 5040°$

Exam tip

The interior angle and exterior angle always lie on a straight line.

Remember

Rearrange $E = \frac{360}{n}$ in the form $n = \frac{360}{E}$ to find the number of sides.

Watch out!

The hexagon isn't regular so the angles won't all be the same!

Exam tip

First find the sum of interior angles using $180 \times (n-2)$. Then form an equation in terms of x.

Exam tip

Learn the names of these common polygons:
- Pentagon – 5 sides
- Hexagon – 6 sides
- Heptagon – 7 sides
- Octagon – 8 sides
- Decagon – 10 sides

Exam-style questions

1. The exterior angle of a regular polygon is 15°
   - a Work out the number of sides of the polygon. [1]
   - b Work out the sum of the interior angles of the polygon. [2]

2. A regular polygon has n sides.

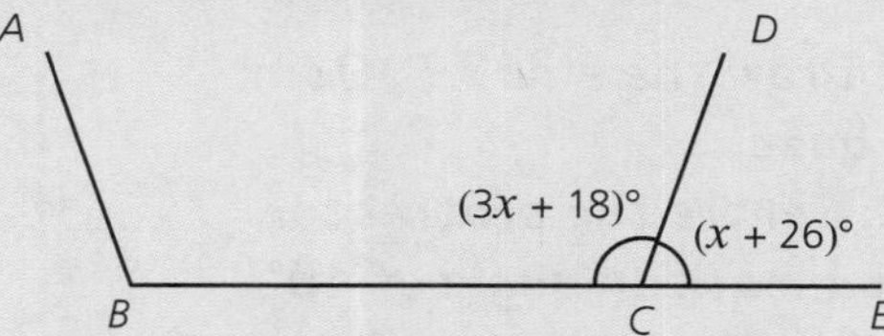

Each interior angle is $(3x + 18)°$
Each exterior angle is $(x + 26)°$
Work out the value of n. [3]

Short answers on page 140
Full worked solutions online

CHECKED ANSWERS ONLINE

Constructions

REVISED

Key facts

To carry out constructions you need

- a sharp pencil
- a clear ruler
- compasses that don't go loose when you use them
- a protractor.

Make sure you

- measure as accurately as you can
- leave all your construction marks on your diagram.

You need to know how to draw each of the six constructions shown in the examples.

The examples are **not** to scale, but they show you how to carry out each type of construction.

Worked examples

Construct a triangle given 3 sides

ABC is an isosceles triangle. $AB = 10$ cm, $AC = BC = 7$ cm. Using a ruler and compasses construct triangle *ABC* showing all construction lines.

Solution

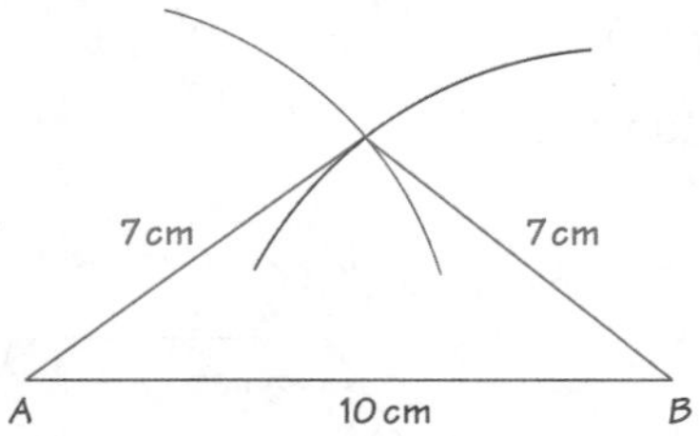

Draw *AB* as a base.

Open compasses with radius 7 cm, draw an arc with centre *B*.

Repeat with an arc with centre *A*.

Join each end of *AB* to where the arcs meet.

7 cm 7 cm
A 10 cm B

Construct a triangle with two sides and the angle between them

ABC is a triangle. $AB = 9$ cm, $AC = 4$ cm and angle $BAC = 55°$. Make an accurate drawing of triangle *ABC*.

Solution

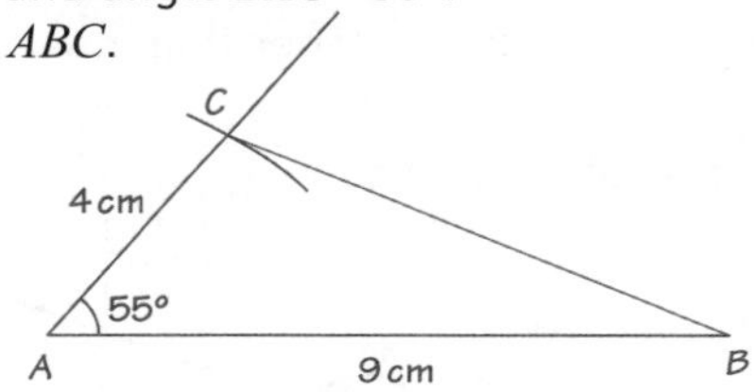

Draw *AB*, 9 cm, as the base.

At *A* measure an angle of 55° and draw a line through this point.

Using compasses draw an arc of 4 cm from *A* to locate *C*.

Then join *B* to *C* to complete the construction.

C
4 cm
55°
A 9 cm B

Construct a triangle given two angles and the included side

Draw triangle *ABC*, where $AB = 9$ cm, angle $BAC = 45°$ and angle $ABC = 60°$

Solution

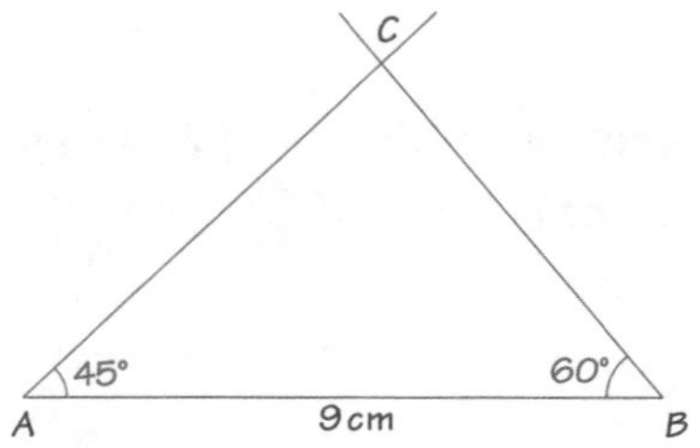

Step 1: Draw the side *AB*, 9 cm, as the base.

Step 2: Centre the protractor on *A* and mark an angle of 45°.
Draw a line from *A* through this point.

Step 3: Centre the protractor on *B* and mark an angle of 60°
Draw a line from *B* through this point.

Step 4: Label the point of intersection as *C*.

C
45° 60°
A 9 cm B

Exam tip

Don't rub out your construction arcs.

Remember

Never try to just measure lines instead of constructing them otherwise you may lose marks.

Exam tip

Draw large arcs so that they are 'long' enough to intersect each other.

Construct a triangle with two given sides and an angle not between them

Draw a triangle ABC given $AB = 10$ cm, $BC = 8$ cm and angle $CAB = 50°$

Solution

Step 1: Draw the longest side AB, 10 cm, as the base.

Step 2: Construct a line from A at an angle of 50°

Step 3: Open compasses to a radius of 8 cm, drawing an arc from B.

Step 4: The arc will intersect the original line from A at two points, giving two ways of completing the construction.

Construct a perpendicular bisector of a line segment

Construct the perpendicular bisector of the line segment AB as shown.

Solution

Step 1: Open compasses a little over half the length of AB.

Construct arcs from A and B.

Step 2: Join the two points where the arcs meet.

Construct an angle bisector

Use a ruler and compasses to construct the angle bisector of angle A.

Solution

Step 1: Using compasses, draw an arc from centre A, passing through both lines at X and Y.

Step 2: Without adjusting the compasses construct two further arcs with centres at X and Y.

Step 3: Complete the construction by drawing a straight line from A to where the arcs intersect. The original angle has now been bisected.

Exam-style questions

1 ABC is a triangle.
Showing all your construction lines,

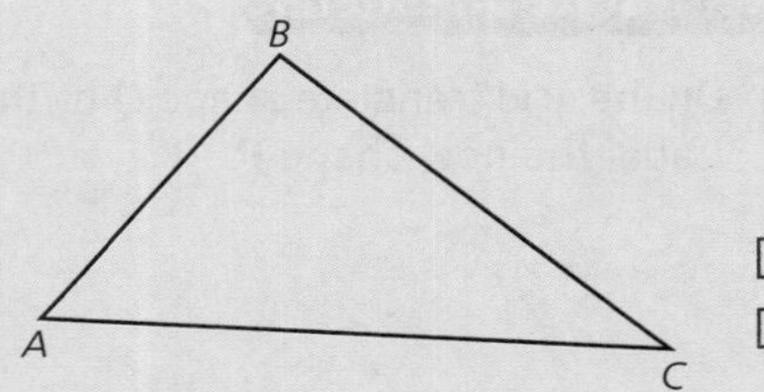

- **a** construct the bisector of angle ACB [2]
- **b** construct the perpendicular bisector of AB. [2]

2 Triangle ABC has $AB = 8$ cm, $BC = 5$ cm and angle $CAB = 35°$.
Construct the two possible triangles with these measurements. [2]

3 Construct triangle ABC where $AB = 12$ cm, $BC = 9$ cm and angle $ABC = 57°$ [2]

Short answers on page 140

Full worked solutions online

CHECKED ANSWERS ONLINE

Translations

REVISED

Key facts

1 A **translation** moves a shape by a vector $\begin{pmatrix} x \\ y \end{pmatrix}$

Example: $\begin{pmatrix} 8 \\ 6 \end{pmatrix}$ means a translation of 8 units to the right and 6 units up.

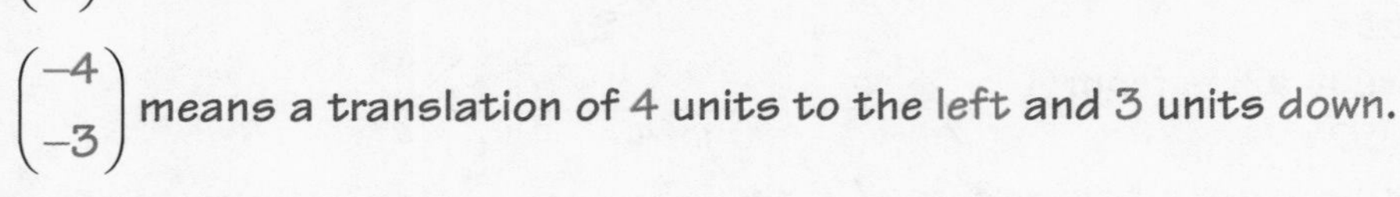
$\begin{pmatrix} -4 \\ -3 \end{pmatrix}$ means a translation of 4 units to the left and 3 units down.

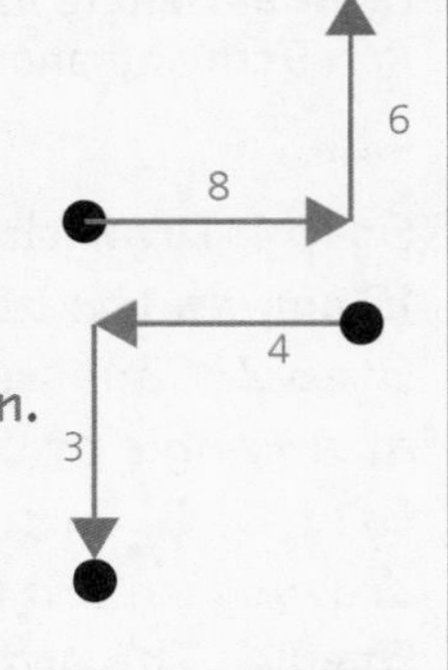

2 **Congruent shapes** are exactly the same shape and size.
A shape is **congruent** under translation.

Worked examples

Describing a translation

The diagram shows shapes **S**, **T** and **U** on a grid.

Describe fully the single transformation that:

a Maps **S** onto **T**.

b Maps **S** onto **U**.

Solution

To get from S to T, you go 1 unit left and 2 units down.

So this is a translation of $\begin{pmatrix} -1 \\ -2 \end{pmatrix}$

To get from S to U, you go 2 units right and 2 units up.

So this is a translation of $\begin{pmatrix} 2 \\ 2 \end{pmatrix}$

Using a translation vector

Translate shape **A** by the vector $\begin{pmatrix} 6 \\ -5 \end{pmatrix}$

Label the new shape **B**.

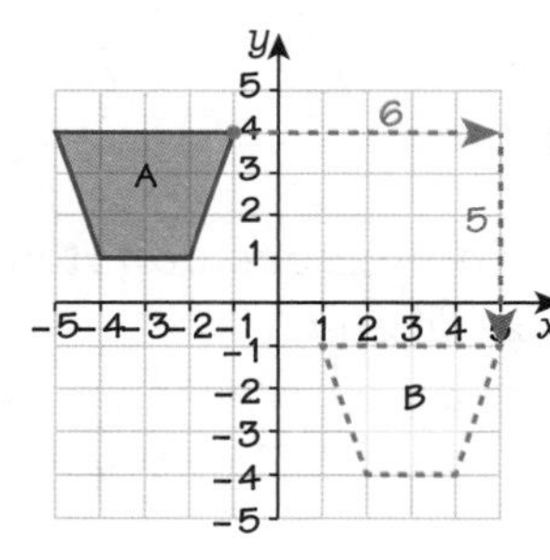

Solution

Pick a corner on shape A and go 6 units right and 5 units down.

Remember

A translation is one type of transformation.

Exam tip

The correct mathematical term to describe how a shape is moved by a vector is **translation** *not* **move** or **slide**.

Watch out!

Don't use coordinates to describe a translation. So write $\begin{pmatrix} -1 \\ -2 \end{pmatrix}$ and not $(-1,-2)$.

Exam-style questions

1 On the grid translate shape **Q** by the vector $\begin{pmatrix} -5 \\ 2 \end{pmatrix}$. Label the new shape **R**. [1]

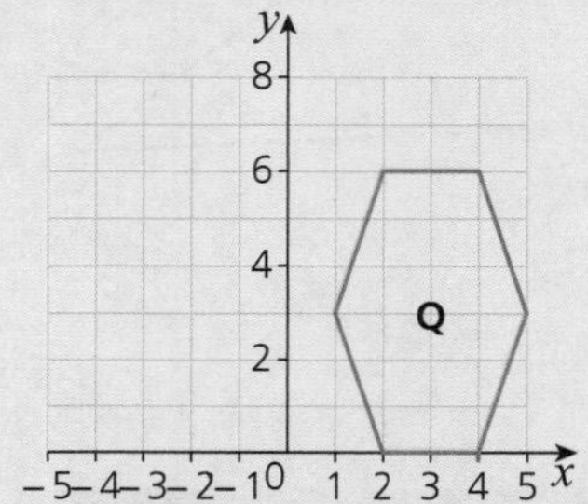

2 The diagram shows shapes **M** and **N**.

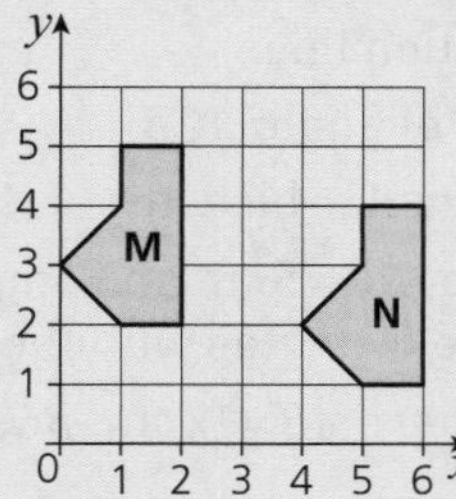

Describe fully the transformation that maps shape **M** onto shape **N**. [2]

Short answers on page 141

Full worked solutions online

CHECKED ANSWERS ONLINE

Reflections

REVISED

Key facts

1. A **reflection** is a mirror image of a 2D shape.
2. Each point of an **object** is the same distance away from the mirror line as the **image**.
3. A **mirror line** can be horizontal, vertical or at an angle.
4. A shape is **congruent** under a reflection.

Worked examples

Reflections in the x- and y-axes

Shape **A** is drawn on the grid.

a Reflect **A** in the y-axis and label this **B**.

b Reflect **B** in the x-axis and label this **C**.

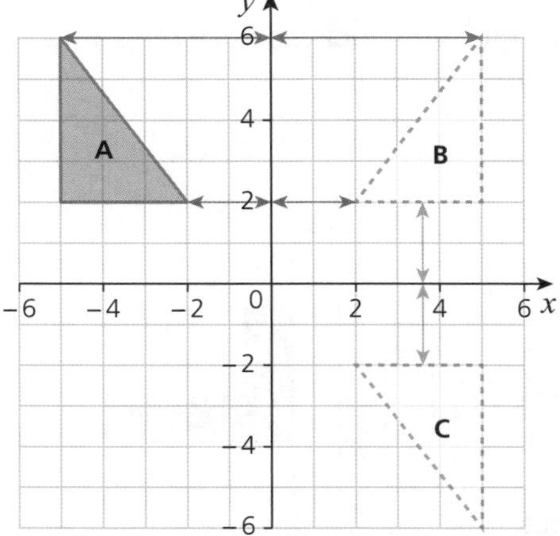

Solution

Each vertex of the image must be the same distance away from the mirror line as the object.

Mirror lines at an angle

Shape **P** is drawn on the grid.

Reflect shape **P** in the line $y = -x$.

Label the new shape **Q**.

Solution

Reflecting in the line $y = -x$ reverses the coordinates and also the signs:

$(-6, -3) \rightarrow (3, 6)$,

$(-6, -5) \rightarrow (5, 6)$;

$(-4, -2) \rightarrow (2, 4)$; $(-4, -6) \rightarrow (6, 4)$ and $(-2, -4) \rightarrow (4, 2)$

Watch out!

Make sure you reflect the shape in the correct mirror line.

Remember

A horizontal line has an equation of the form $y = a$. A vertical line has an equation of the form $x = b$.

Exam tip

Always draw the mirror line. Doing this can gain you a mark!

Exam-style questions

1 On the grid below reflect shape **P** in the line $y = x$ [2]

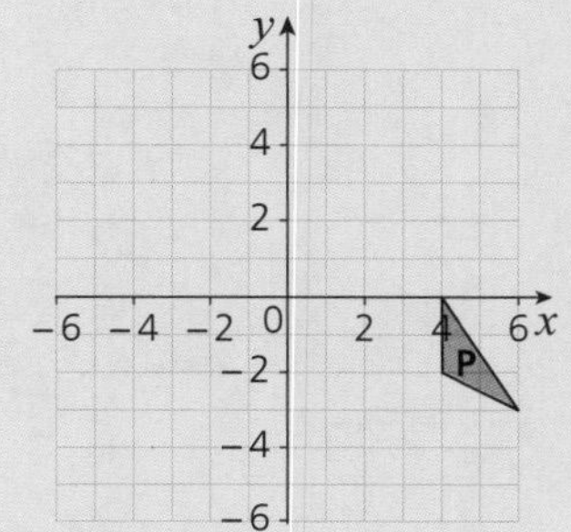

2 The diagram shows a trapezium and a horizontal line AB.

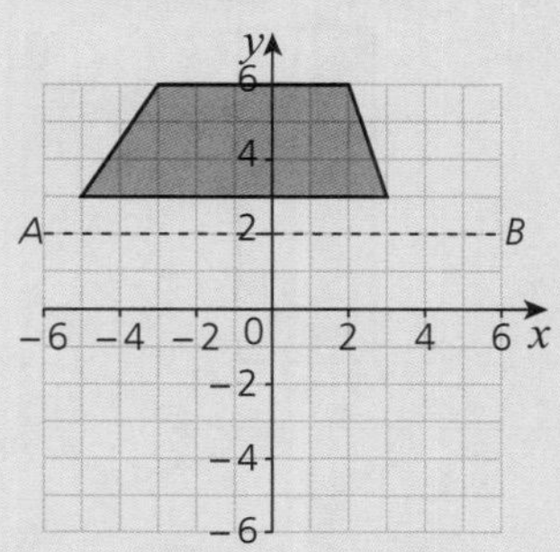

a Write down the equation of the line AB. [1]

b Reflect the trapezium in the line AB. [2]

Short answers on page 141

Full worked solutions online

CHECKED ANSWERS ONLINE

Geometry and transformations

Rotations

REVISED

Key facts

1. A **rotation** will turn an object to a different position about a fixed point called the **centre of rotation**.
2. To describe a rotation you need to state the:
   - angle of rotation
   - direction of rotation (clockwise or anticlockwise)
   - centre of rotation.
3. Rotations of 180° clockwise and 180° anticlockwise are the same.
4. Shapes are congruent under a rotation.

Remember

You can use tracing paper in the exam. Make sure you ask for some and use it to check your answer. Remember that a rotation of 270° clockwise is the same as 90° anticlockwise.

Worked examples

Performing a rotation

On the grid, rotate shape **L** through 270° clockwise about the point (1, 0). Label the new shape **M**.

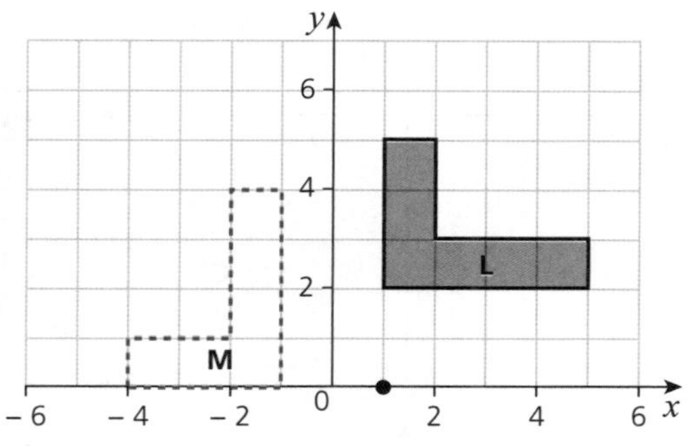

Solution

Use tracing paper to trace the **object** L. Fix the centre of rotation with a pencil point. **Rotate** through 270° clockwise. Draw the **image** onto the grid.

Describing a rotation

Describe fully the single transformation which maps shape **Q** onto shape **P**.

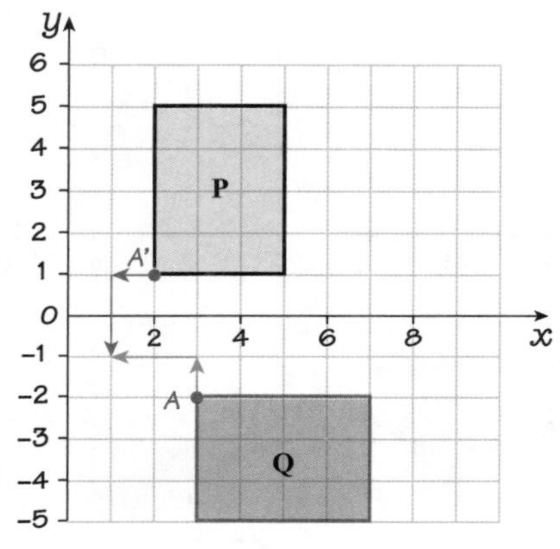

Solution

A rotation of 90° anticlockwise about centre (1, –1).

Watch out!

Don't use the word *turn* to describe a rotation.

Exam tip

You can use tracing paper to help you find the centre of rotation.
Or you can pick two corresponding corners, and by eye find a point they are an equal distance from. A to (1, –1) is 1 square up and 2 across. A' to (1, –1) is 1 square across and 2 down. So (1, –1) is the centre of rotation.

Exam tip

Writing the word *rotation* alone will score a mark!

Exam-style questions

1 Describe fully the single transformation that maps rectangle **A** onto rectangle **B**. [3]

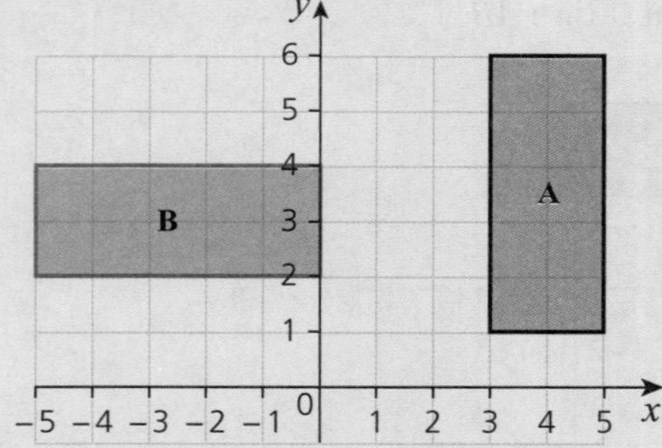

2 On the grid, rotate the triangle **T** 270° anticlockwise centre (2, 2) [2]

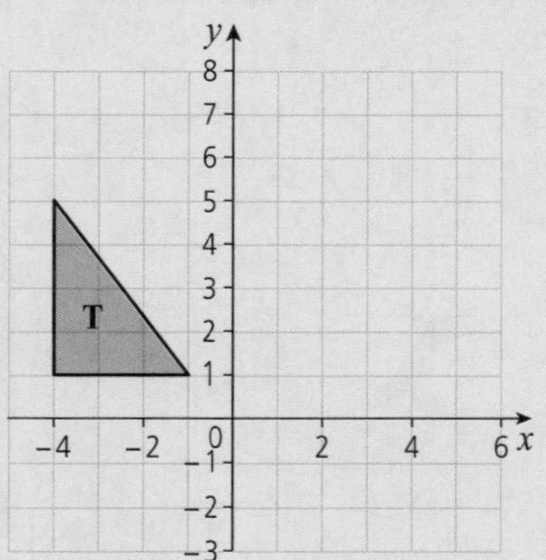

Short answers on page 141

Full worked solutions online

CHECKED ANSWERS ONLINE

Enlargements

REVISED

Key facts

1. **Enlargements** change the size of a 2D shape.
2. They have a centre of enlargement and scale factor.
   - A **scale factor** greater than 1: the shape gets bigger.
   - A **scale factor** between 0 and 1: the shape gets smaller.
   - Scale factor = $\frac{\text{new length}}{\text{old length}}$
3. Shapes are **similar** under enlargement.

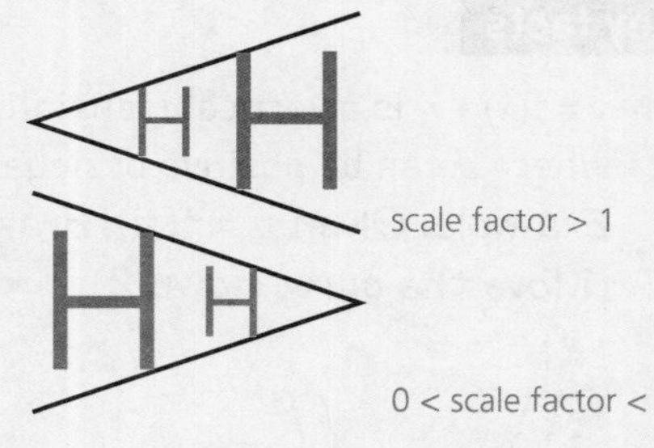

Worked examples

A transformation as an enlargement

Describe fully the single transformation which maps triangle **P** onto triangle **Q**.

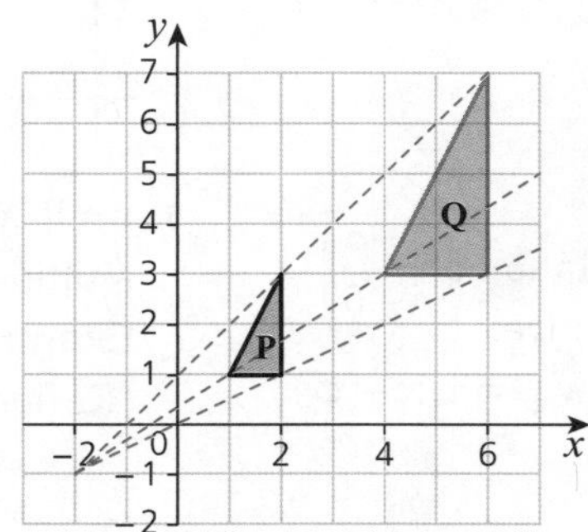

Solution

By counting squares, shape Q is twice the width and height of shape P.

So this is an enlargement, scale factor 2, centre (−2, −1)

Fractional scale factor and counting squares

Enlarge shape **T** by scale factor $\frac{1}{3}$, centre (−2, 1)

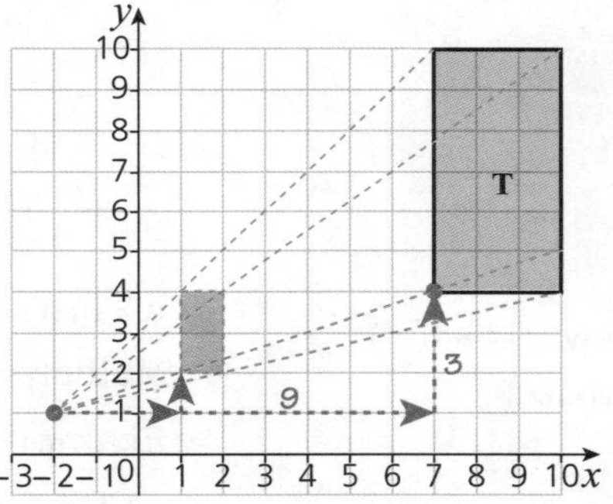

Solution

From (−2, 1) to a corner of the shape T is, by counting, 9 squares right and 3 squares up.

As the scale factor is $\frac{1}{3}$ mark one corner of the image shape as 3 squares right $\left(\frac{1}{3} \times 9 = 3\right)$ and 1 square up $\left(\frac{1}{3} \times 3 = 1\right)$.

T is 3 squares by 6 squares, so the image is 1 square by 2 squares.

Remember

Three elements are needed to describe the transformation:

- enlargement
- scale factor
- centre.

Exam tip

Draw rays from **Q** to each corner of the original shape **P**. The rays cross at the centre of enlargement. Make sure all your rays are long enough so they meet at the same point.

Exam-style questions

1 Describe fully the single transformation that maps shape **A** onto shape **B**. [3]

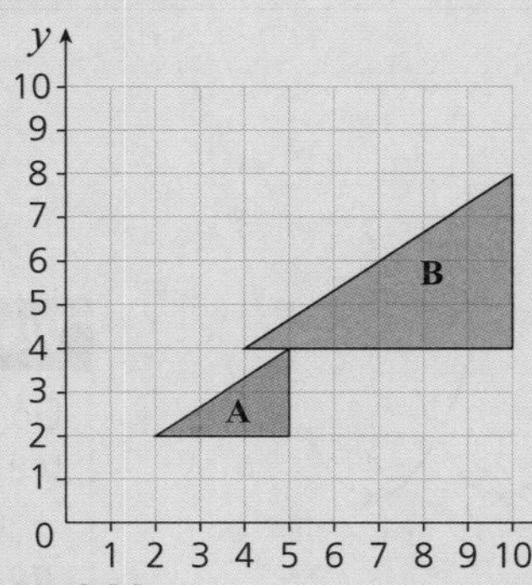

2 Enlarge shape **Q** by scale factor $\frac{1}{2}$, centre A. Label the new shape **R**. [2]

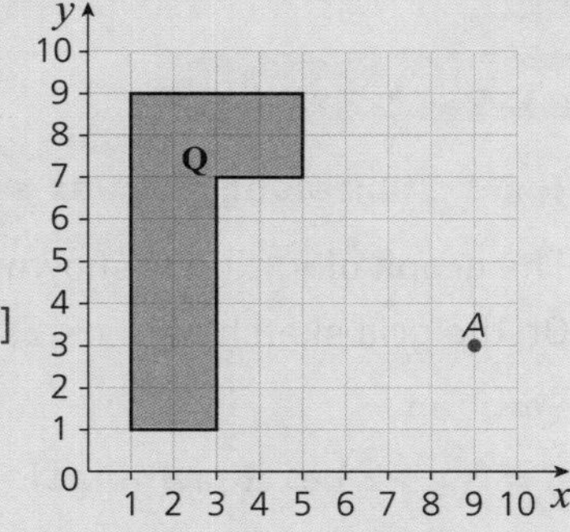

Short answers on page 141

Full worked solutions online

CHECKED ANSWERS ONLINE

Geometry and transformations

Graphs and transformations

REVISED

Key facts

1 $y = f(x) + a$ is a vertical translation of $\begin{pmatrix} 0 \\ a \end{pmatrix}$ where a can be positive or negative.
Example: Given $y = f(x)$, draw $y = f(x) - 2$
(Move the curve down 2)

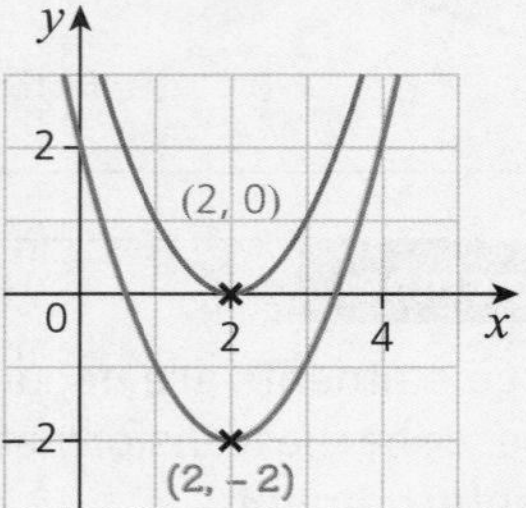

2 $y = f(x + a)$ is a horizontal translation of $\begin{pmatrix} -a \\ 0 \end{pmatrix}$ where a can be positive or negative.
Example: Given $y = f(x)$, draw $y = f(x - 5)$
(Move curve 5 right.)

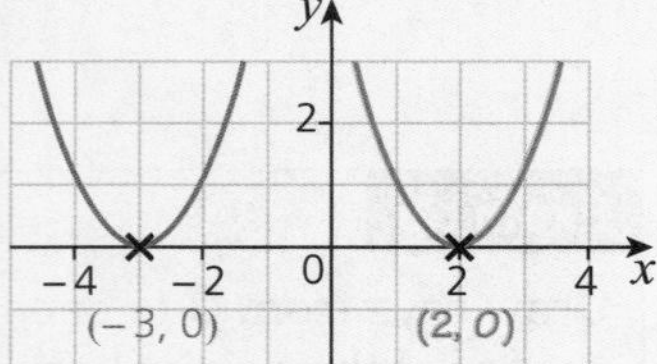

3 $y = -f(x)$ is a reflection in the x-axis.
Multiply y-coordinates by −1
Example: Given $y = f(x)$, draw $y = -f(x)$
(Reflect the curve in the x-axis.)

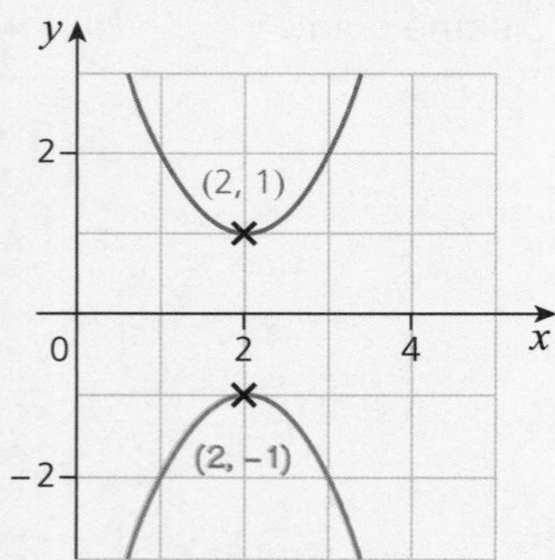

4 $y = af(x)$ is a vertical stretch scale factor a.
Multiply y-coordinates only by a.
Example: Given $y = f(x)$, draw $y = 2f(x)$
(All y-coordinates are doubled.)

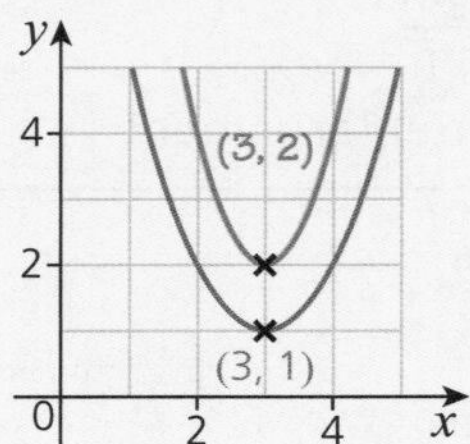

5 $y = f(-x)$ is a reflection in the y-axis.
Multiply x-coordinates by −1
Example: Given $y = f(x)$, draw $y = f(-x)$
(Reflect the curve in the y-axis.)

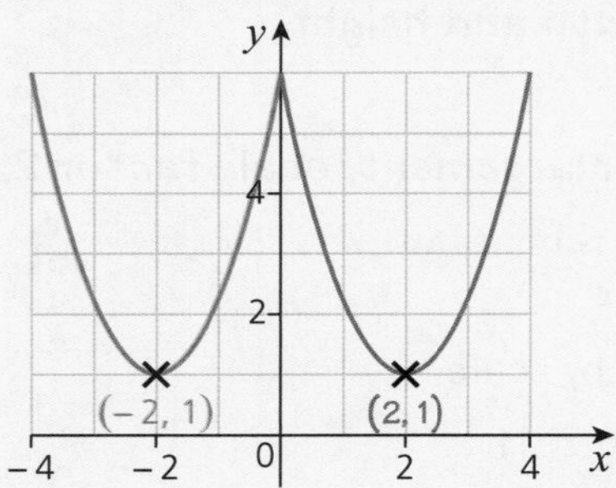

6 $y = f(ax)$ is a horizontal stretch scale factor $\frac{1}{a}$
Multiply x-coordinates only by $\frac{1}{a}$
Example: Given $y = f(x)$, draw $y = f(2x)$
(All x-coordinates are halved.)

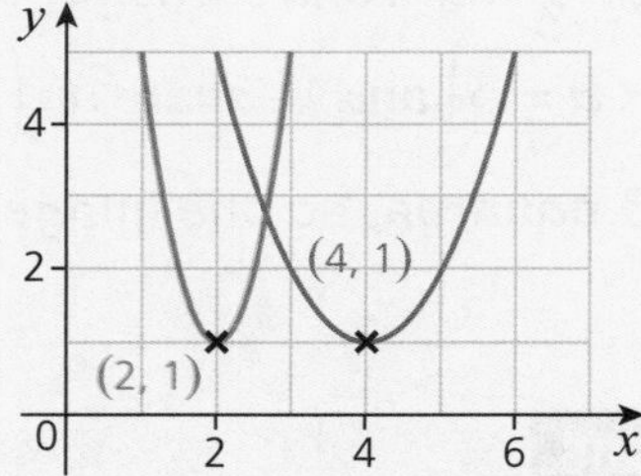

Worked examples

Translation along the x-axis

The graph of $y = f(x)$ is shown.
On the grid sketch $y = f(x + 2)$

Solution

$y = f(x + 2)$ is a translation of $\begin{pmatrix} -2 \\ 0 \end{pmatrix}$

Move the curve left 2 so $(4, -1) \rightarrow (2, -1)$

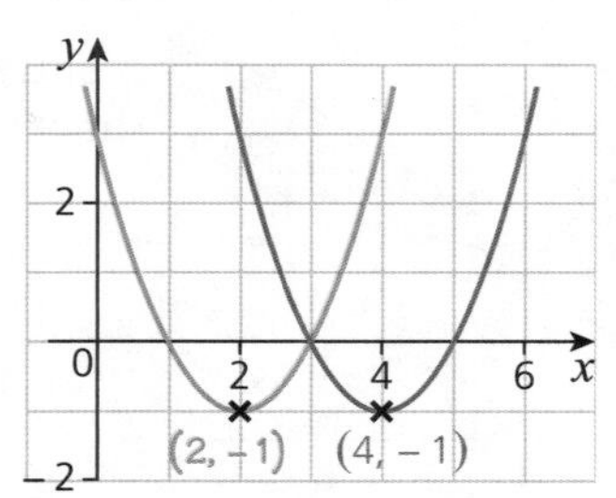

Watch out!

Do the opposite with translation in the x-axis. '+' in the brackets means '*go left*'. '−' in the brackets means '*go right*'.

Transformation of $af(x)$

The diagram shows a sketch of $y = \cos x$

On the same diagram sketch $y = \frac{1}{2}\cos x$

Solution

*Divide each y **coordinate** by 2*

$y = \cos x$ has max/min values at ±1

$y = \frac{1}{2}\cos x$ will have max/min values at $\pm\frac{1}{2}$

Transformation of $-g(x)$ and $g(-x)$

The graph of $y = g(x)$ is is shown.
Draw the graph of $y = -g(x)$
Draw the graph of $y = g(-x)$

Solution

$y = -g(x)$ is a reflection in the x-axis.

$y = g(-x)$ is a reflection in the x-axis.

Working with turning points

The curve $y = f(x)$ has a minimum turning point at (–2, –4). Write down the turning point of the curve with equation

a $y = f(x + 4)$
b $y = 3f(x)$
c $y = f(2x)$

Solution

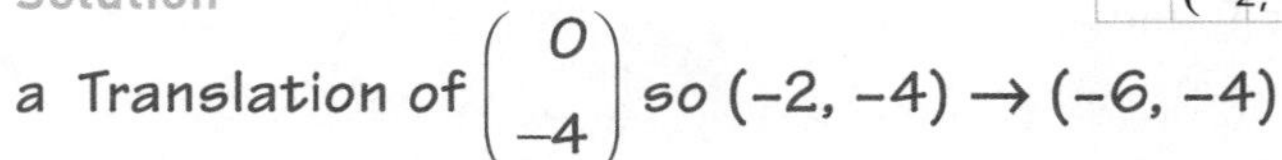

a Translation of $\begin{pmatrix} 0 \\ -4 \end{pmatrix}$ so (–2, –4) → (–6, –4)

b Multiply the y coordinate by 3 so (–2, –4) → (–2, –12)

c Divide the x coordinate by 2 so (–2, –4) → (–1, –4)

Exam tip

Look for points where the graph passes through whole number coordinates and transform these points.

Remember

A transformation of $y = af(x)$ means the x coordinates stay the same.

Remember

When reflecting in the x-axis multiply the ***y* coordinate** by –1
When reflecting in the y-axis multiply the ***x* coordinate** by –1

Exam-style questions

1 The diagram shows a sketch of $y = \cos x°$

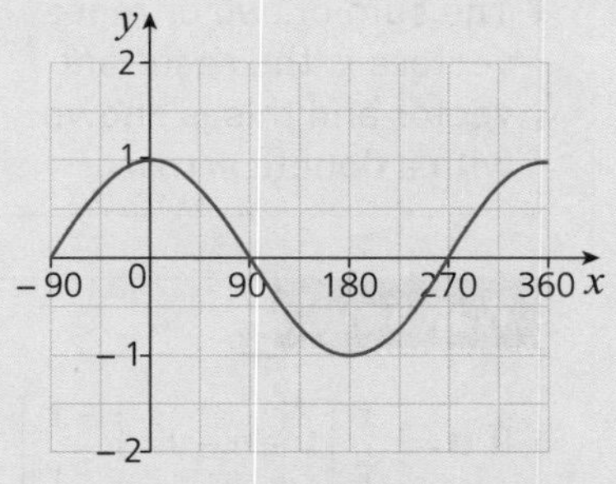

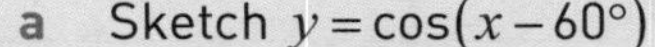

a Sketch $y = \cos(x - 60°)$ [2]
b Sketch $y = \cos x - 4$ [2]
c Sketch $y = -3\cos x$ [2]

2 The maximum point of a curve $y = f(x)$ is (2, 1)

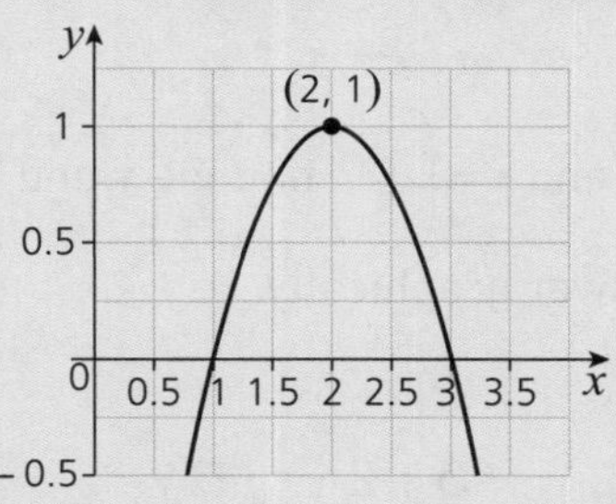

Find the coordinates of the maximum point when:

a $y = f(x + 3) - 2$ [2]
b $y = 3 + f(x + 1)$ [2]

Short answers on pages 141–142

Full worked solutions online

CHECKED ANSWERS ONLINE

Vectors

REVISED

Key facts

1 A **vector** is a quantity with both size (magnitude) and direction.
A **scalar** is a quantity that has size only.

2 A vector can be written as a column vector.

Example: $\begin{pmatrix} 3 \\ -2 \end{pmatrix}$ means 2 units right and 2 units down. $\begin{pmatrix} -2 \\ 4 \end{pmatrix}$ means 2 units left and 4 units up.

3 Vectors can be represented by a straight line.
The movement from O to A can be represented by **a** (as seen in books and exams) or $\underline{a}$ (underline vectors when writing by hand) or $\overrightarrow{OA}$ (vector from point O to point A).
$\overrightarrow{AB} = \overrightarrow{OB} - \overrightarrow{OA} = \mathbf{b} - \mathbf{a}$

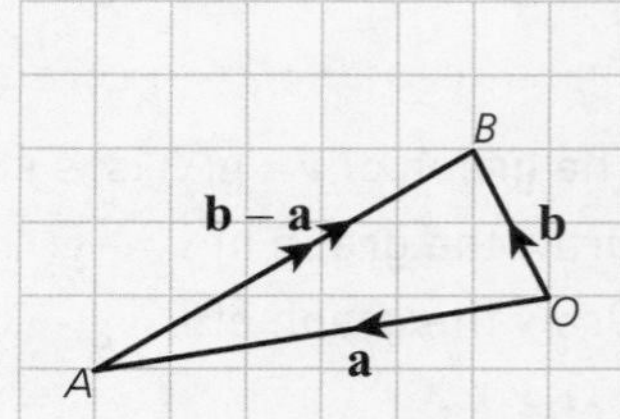

4 Addition and subtraction of vectors is done using simple arithmetic.

Example: $\begin{pmatrix} 6 \\ 1 \end{pmatrix} + \begin{pmatrix} 4 \\ 2 \end{pmatrix} = \begin{pmatrix} 6+4 \\ 1+2 \end{pmatrix} = \begin{pmatrix} 10 \\ 3 \end{pmatrix}$

$\begin{pmatrix} 3 \\ 7 \end{pmatrix} - \begin{pmatrix} 4 \\ -6 \end{pmatrix} = \begin{pmatrix} 3-4 \\ 7--6 \end{pmatrix} = \begin{pmatrix} -1 \\ 13 \end{pmatrix}$

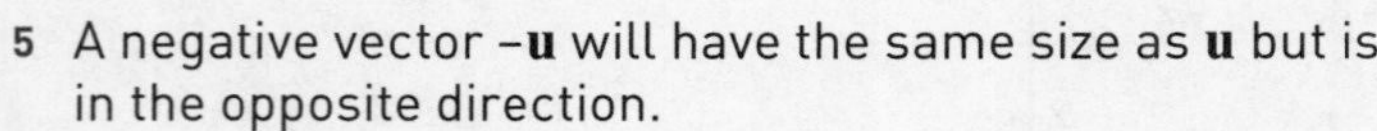

The addition of vectors can be shown using a diagram.
The sum of two vectors is called the **resultant** and this is shown with a double arrow.

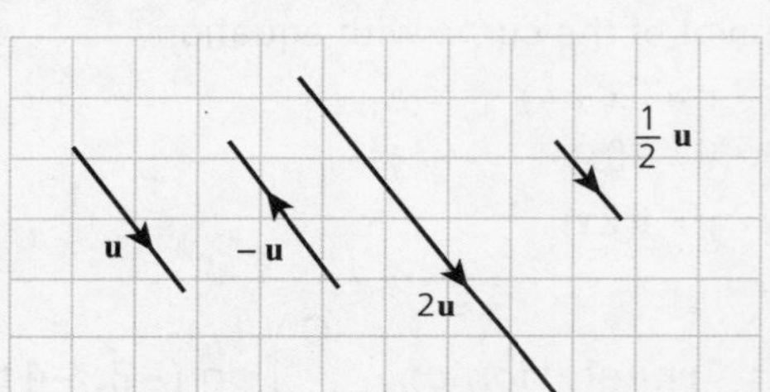

5 A negative vector $-\mathbf{u}$ will have the same size as $\mathbf{u}$ but is in the opposite direction.

- $\mathbf{u}$ and $-\mathbf{u}$ are parallel vectors
- **scalar** multiples of vectors are **parallel**.

6 The **magnitude** of a vector is its length and this is found using Pythagoras' theorem.

Example: $\mathbf{a} = \begin{pmatrix} 3 \\ 4 \end{pmatrix}$

So magnitude of $\mathbf{a} = \sqrt{3^2 + 4^2} = \sqrt{9+16} = \sqrt{25} = 5$

Worked examples

Working with column vectors

Given that $\mathbf{u} = \begin{pmatrix} 3 \\ 7 \end{pmatrix}$ and $\mathbf{v} = \begin{pmatrix} -2 \\ 6 \end{pmatrix}$ find $\mathbf{u} - \mathbf{v}$ and illustrate your answer graphically.

Solution

$\underline{u} - \underline{v} = \begin{pmatrix} 3 \\ 7 \end{pmatrix} - \begin{pmatrix} -2 \\ 6 \end{pmatrix} = \begin{pmatrix} 3+2 \\ 7-6 \end{pmatrix} = \begin{pmatrix} 5 \\ 1 \end{pmatrix}$

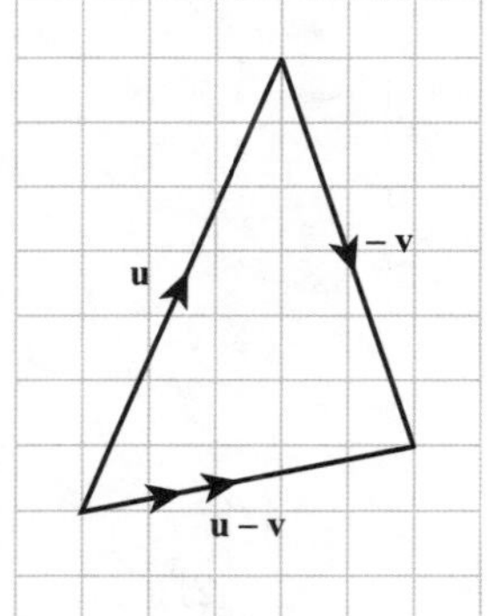

Remember

The sum of two or more vectors is the resultant vector and this is shown with a double arrow.

Watch out!

If $\mathbf{u} = \begin{pmatrix} x \\ y \end{pmatrix}$ then $-\mathbf{u} = \begin{pmatrix} -x \\ -y \end{pmatrix}$

Finding a column vector

A, B and C are points such that:

$\overrightarrow{AC} = \begin{pmatrix} 2 \\ -3 \end{pmatrix}$ and $\overrightarrow{BC} = \begin{pmatrix} 9 \\ 8 \end{pmatrix}$

Find $\overrightarrow{BA}$ as a column vector.

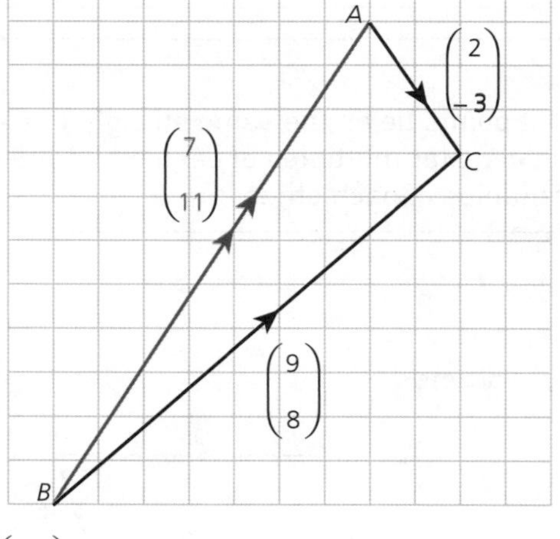

Solution

$\overrightarrow{BA} = \overrightarrow{BC} + \overrightarrow{CA}$

$= \begin{pmatrix} 9 \\ 8 \end{pmatrix} + \begin{pmatrix} -2 \\ 3 \end{pmatrix} = \begin{pmatrix} 7 \\ 11 \end{pmatrix}$

Watch out!

Question gives $\overrightarrow{AC}$ but you need to use $\overrightarrow{CA}$.

Working with vectors

Given $\mathbf{m} = \begin{pmatrix} 6 \\ -3 \end{pmatrix}$, $\mathbf{n} = \begin{pmatrix} 2 \\ 9 \end{pmatrix}$ and $\mathbf{p} = \begin{pmatrix} -1 \\ -2 \end{pmatrix}$

a Write $3\mathbf{m}$ as a column vector.
b Write $2\mathbf{n} - \mathbf{p}$ as a column vector.
c Find the magnitude of $\mathbf{n}$ giving your answer as a surd.

Solution

a $3\underline{m} = 3\begin{pmatrix} 6 \\ -3 \end{pmatrix} = \begin{pmatrix} 3\times 6 \\ 3\times -3 \end{pmatrix} = \begin{pmatrix} 18 \\ -9 \end{pmatrix}$

b $2\underline{n} - \underline{p} = 2\begin{pmatrix} 2 \\ 9 \end{pmatrix} - \begin{pmatrix} -1 \\ -2 \end{pmatrix} = \begin{pmatrix} 4 \\ 18 \end{pmatrix} - \begin{pmatrix} -1 \\ -2 \end{pmatrix} = \begin{pmatrix} 4 - -1 \\ 18 - -2 \end{pmatrix} = \begin{pmatrix} 5 \\ 20 \end{pmatrix}$

c Magnitude of $\underline{n} = \sqrt{2^2 + 9^2} = \sqrt{4 + 81} = \sqrt{85}$

Remember

The magnitude of a vector is found using Pythagoras' theorem.

Finding missing vector components

Given $3\begin{pmatrix} x \\ 7 \end{pmatrix} + 2\begin{pmatrix} 4 \\ y \end{pmatrix} = \begin{pmatrix} 26 \\ 11 \end{pmatrix}$, find the value of x and the value of y.

Solution

$\begin{pmatrix} 3x \\ 21 \end{pmatrix} + \begin{pmatrix} 8 \\ 2y \end{pmatrix} = \begin{pmatrix} 26 \\ 11 \end{pmatrix}$

$3x + 8 = 26$ $\quad$ $21 + 2y = 11$

So: $3x = 18$ and: $2y = -10$

$x = 6$ $\quad$ $y = -5$

Exam-style questions

1 A, C and D are points such that $\overrightarrow{AC} = \begin{pmatrix} 4 \\ -10 \end{pmatrix}$ and $\overrightarrow{DC} = \begin{pmatrix} 10 \\ 12 \end{pmatrix}$

Find $\overrightarrow{DA}$ as a column vector. [2]

2 Given $\overrightarrow{ST} = \begin{pmatrix} 11 \\ -7 \end{pmatrix}$ find the magnitude of $\overrightarrow{ST}$

Give your answer as a surd. [2]

Short answers on page 142
Full worked solutions online

CHECKED ANSWERS ONLINE

Problem solving with vectors

REVISED

Key facts

1. Vectors can be used to show that points lie on the same straight line.
2. Vectors on a straight line must be scalar multiples of each other and share a common point.
3. Vectors that are parallel are multiples of each other.

Worked examples

Using vectors to solve geometric problems

ABCD is a parallelogram.
$\overrightarrow{AC} = 9\mathbf{b}$ and $\overrightarrow{BM} = 4\mathbf{a}$
M is the midpoint of *BC*.
X is the point on *CD* such that $CX : XD = 1 : 2$
Y is the point on *AD* such that $AY : YD = 3 : 1$
Find in terms of **a** and **b**

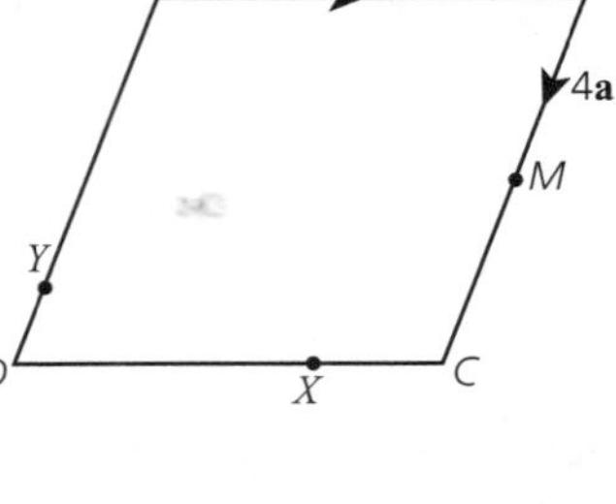

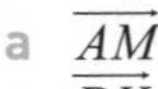

a $\overrightarrow{AM}$ b $\overrightarrow{AC}$
c $\overrightarrow{DX}$ d $\overrightarrow{AY}$
e $\overrightarrow{XY}$

Solution

a $\overrightarrow{AM} = \overrightarrow{AB} + \overrightarrow{BM} = 9\underline{b} + 4\underline{a}$

b $\overrightarrow{AC} = \overrightarrow{AB} + \overrightarrow{BC} = 9\underline{b} + 8\underline{a}$

c $\overrightarrow{DX} = \frac{2}{3}\overrightarrow{DC} = \frac{2}{3}(9\underline{b}) = 6\underline{b}$

d $\overrightarrow{DY} = \frac{1}{4}\overrightarrow{CB} = \frac{1}{4}(-8\underline{a}) = -2\underline{a}$

e $\overrightarrow{XY} = \overrightarrow{XD} + \overrightarrow{DY} = -6\underline{b} - 2\underline{a}$

Using vectors to prove a straight line exists

In the diagram $\overrightarrow{OB} = \mathbf{a}$, $\overrightarrow{AD} = 3\mathbf{b}$, $\overrightarrow{BD} = 2\mathbf{a} - \mathbf{b}$, $\overrightarrow{DC} = 6\mathbf{a} - 11\mathbf{b}$
Show that *OAC* is a straight line.

Solution

For *OAC* to be a straight line $\overrightarrow{OA}$ must be a scalar multiple of $\overrightarrow{OC}$.

$\overrightarrow{OA} = \overrightarrow{OB} + \overrightarrow{BD} + \overrightarrow{DA}$
$= \underline{a} + (2\underline{a} - \underline{b}) - 3\underline{b}$
$= 3\underline{a} - 4\underline{b}$

$\overrightarrow{OC} = \overrightarrow{OB} + \overrightarrow{BD} + \overrightarrow{DC}$
$= \underline{a} + (2\underline{a} - \underline{b}) + (6\underline{a} - 11\underline{b})$
$= 9\underline{a} - 12\underline{b} = 3(3\underline{a} - 4\underline{b})$

$\overrightarrow{OC} = 3\overrightarrow{OA}$ which means $\overrightarrow{OC}$ is a scalar multiple of $\overrightarrow{OA}$ so *OAC* is a straight line.

Using vectors to show a line is not straight

In the diagram *OABC* is a parallelogram.
$\overrightarrow{OA} = \mathbf{a}$ and $\overrightarrow{OC} = \mathbf{b}$
M is the midpoint of $\overrightarrow{BA}$ and *X* is a point on $\overrightarrow{OB}$ such that $OX : XB = 3 : 1$ and *Y* is a point on $\overrightarrow{OA}$ such that $OY : YA = 3 : 1$

Watch out!

Take care when you change a ratio to a fraction.
$AY : YD$ is 3 : 1
So $AY = \frac{3}{4}$ of AD
and $YD = \frac{1}{4}$ of AD
See page 6 for a reminder!

Remember

Think of it as a vector walk. To go from *A* to *M* you have to go from *A* to *B* and then from *B* to *M*. Always follow the direction of the vector.

Remember

OA and *OC* share *O* as a common point so you only need to show that they are multiples of each other.

Watch out!

CA and *CM* share a common point but they are not multiples of each other as *CA* is a multiple of 3**a** – **b** but *CM* is a multiple of 2**a** – **b**

a Is CXM a straight line?

b Is $\overrightarrow{XY}$ parallel to $\overrightarrow{BA}$?

Solution

a For CXM to be a straight line $\overrightarrow{CX}$ must be a multiple of $\overrightarrow{CM}$.

$$\overrightarrow{CX} = \overrightarrow{CO} + \frac{3}{4}\overrightarrow{OB} = -\underline{b} + \frac{3}{4}(\underline{a} + \underline{b}) = -\underline{b} + \frac{3}{4}\underline{a} + \frac{3}{4}\underline{b} = \frac{3}{4}\underline{a} - \frac{1}{4}\underline{b} = \frac{1}{4}(3\underline{a} - \underline{b})$$

$$\overrightarrow{CM} = \overrightarrow{CB} + \frac{1}{2}\overrightarrow{BA} = \underline{a} + \frac{1}{2}(-\underline{b}) = \underline{a} - \frac{1}{2}\underline{b} = \frac{1}{2}(2\underline{a} - \underline{b})$$

$\overrightarrow{CX}$ and $\overrightarrow{CM}$ are not multiples of each other, so CXM is not a straight line.

b
$$\overrightarrow{XY} = \overrightarrow{XO} + \overrightarrow{OY} = \frac{3}{4}\overrightarrow{BO} + \frac{3}{4}\overrightarrow{OA} = \frac{3}{4}(-\underline{a} - \underline{b}) + \frac{3}{4}\underline{a} = -\frac{3}{4}\underline{a} - \frac{3}{4}\underline{b} + \frac{3}{4}\underline{a} = -\frac{3}{4}\underline{b}$$

$\overrightarrow{BA} = -\underline{b}$

$\overrightarrow{XY}$ is a multiple of $\overrightarrow{BA}$ so they are parallel.

Remember

Give a conclusion in your answer.

Exam-style questions

1 The diagram shows a trapezium $ABCD$.

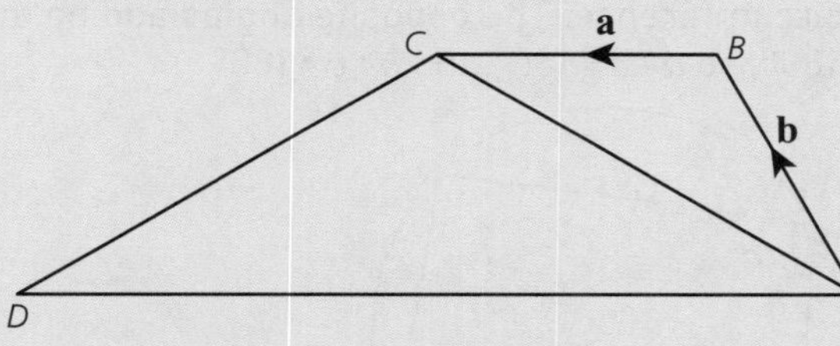

$\overrightarrow{BC}$ is parallel to $\overrightarrow{AD}$ and $\overrightarrow{AD} = 4\overrightarrow{BC}$.

Find the vector $\overrightarrow{CD}$. [2]

2 $LMNO$ is a quadrilateral.

$\overrightarrow{LM} = 3\mathbf{a}$, $\overrightarrow{MN} = 6\mathbf{b}$, $\overrightarrow{NO} = 9\mathbf{c}$

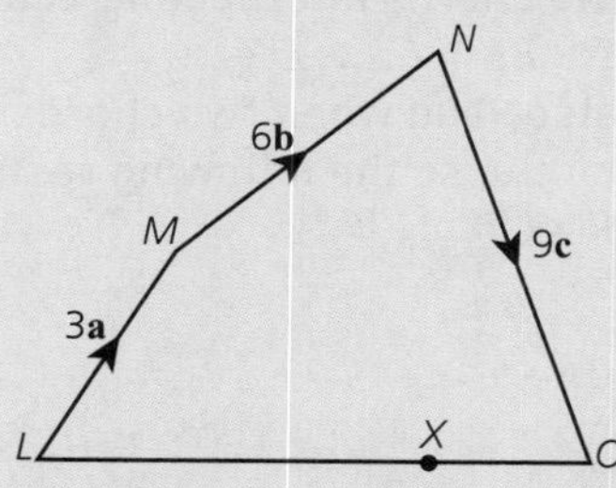

a Find $\overrightarrow{LO}$. [1]

X is on $\overrightarrow{LO}$ such that $LX : XO = 2:1$

b Find $\overrightarrow{LX}$. [2]

c Is $\overrightarrow{MX}$ parallel to $\overrightarrow{NO}$?
Explain your answer. [2]

3 $OABC$ is a parallelogram. $\overrightarrow{OA} = 10\mathbf{a}$ and $\overrightarrow{OC} = 6\mathbf{b}$

X is on CB such that $CB : BX = 2:3$

Y is such that $\overrightarrow{CY} = 2\overrightarrow{AX}$.

Find an expression for $\overrightarrow{OY}$ in terms of $\mathbf{a}$ and $\mathbf{b}$. [3]

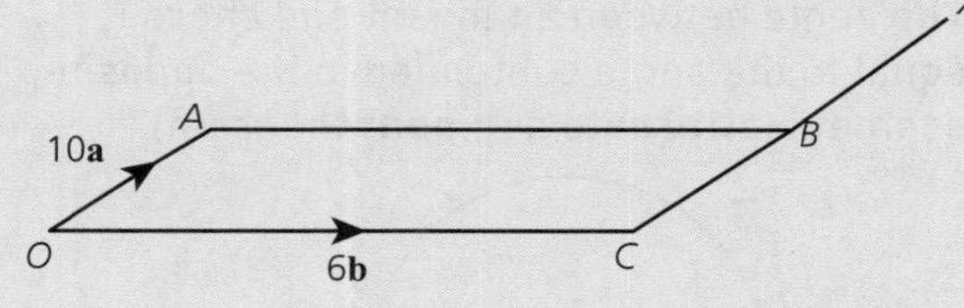

Short answers on page 142

Full worked solutions online

CHECKED ANSWERS ONLINE

Circle theorems

REVISED

Key facts

1 A tangent, AB, is a straight line which touches a circle once. A radius and tangent meet at right angles.

A chord, XY, joins two points on a circumference of a circle. A line drawn from the centre, O, will always bisect a chord.

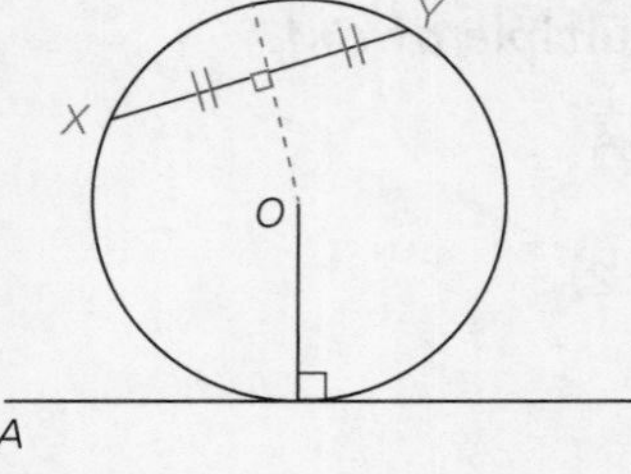

2 The angle at the centre of a circle is twice the angle at the circumference which is subtended by the same arc.

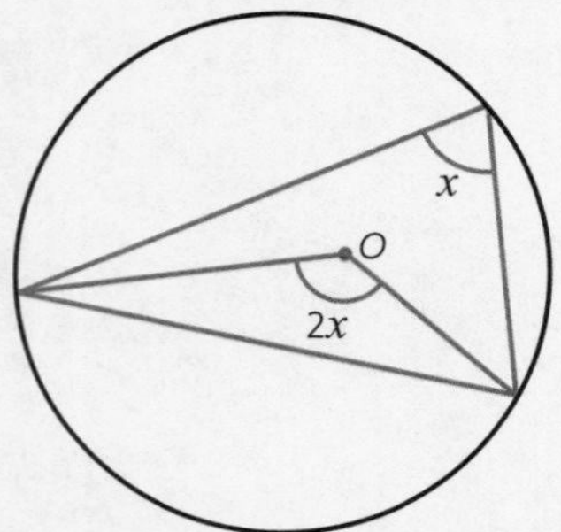

3 Angles subtended by an arc in the same segment of a circle are all equal. Angles at P, Q, R and S are all equal.

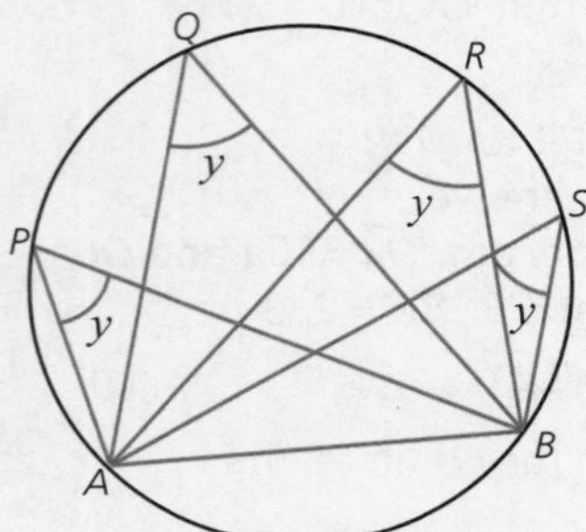

4 The angle between a tangent and chord is equal to the angle subtended in the opposite segment (**alternate segment theorem**).

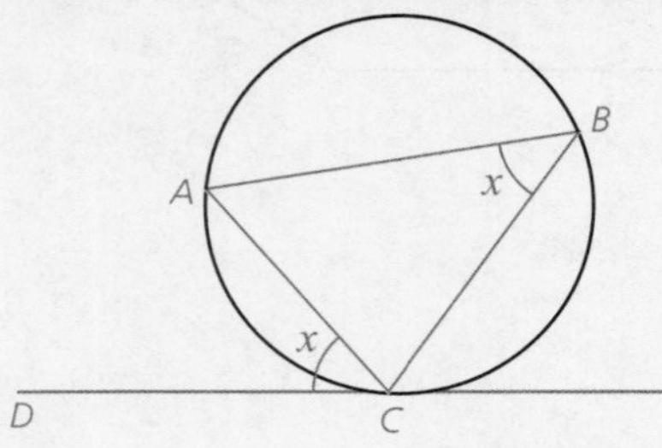

5 Tangents drawn from an outside point are the same length, $ST = SU$.

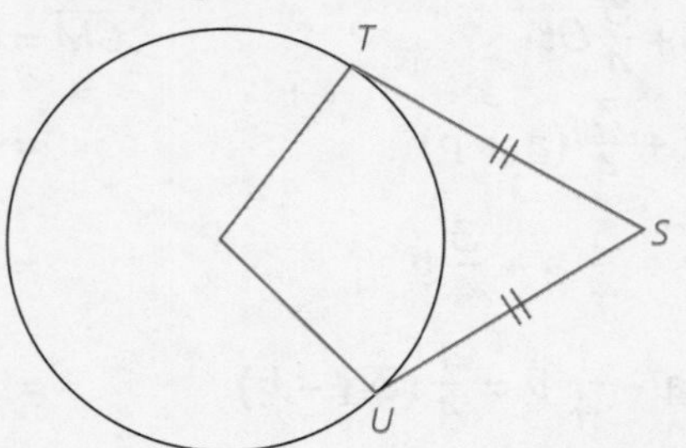

6 The angle at the circumference of a semicircle is always a right angle, where AB is a diameter.

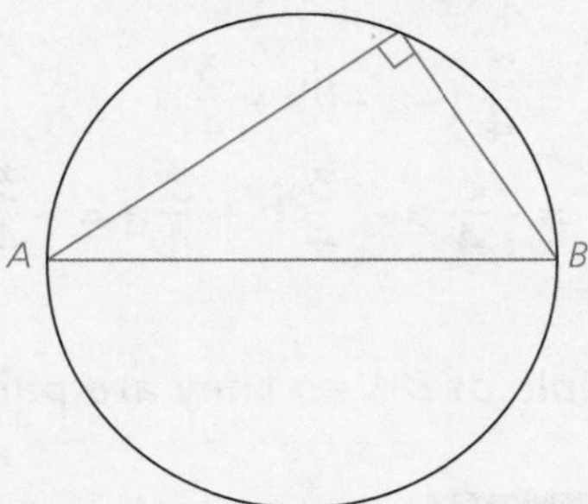

7 A cyclic quadrilateral is a four-sided shape where each of the vertices touches the circumference. The opposite angles add up to 180°, so $a+d=180°$ and $b+c=180°$

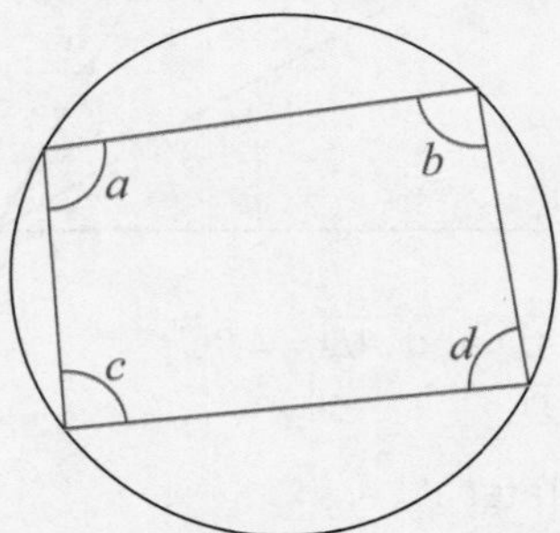

8 AB and CD are two chords intersecting at a point X. Then $AX \times BX = CX \times DX$

The theorem is also valid when two chords cross outside a circle, so the following result applies: $SW \times TW = UW \times VW$

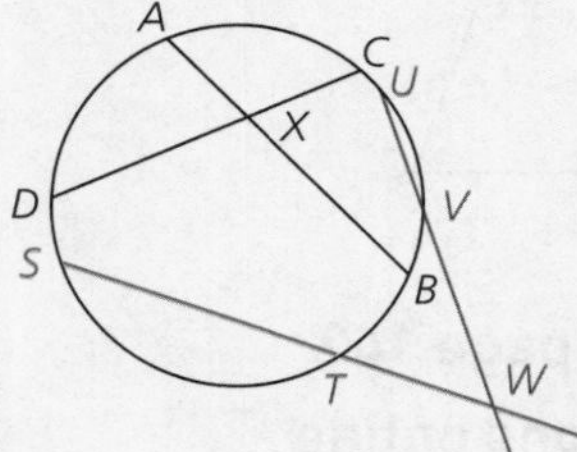

Worked examples

Finding angles at the circumference

A, B and C are points on a circle centre O.
DA and DC are tangents. Angle $ADC = 64°$
Work out the size of angle ABC.

Solution

Angle OAD = Angle $OCD = 90°$ as DA and DC are tangents. Sum of angles in $AOCD = 360°$

Angle $AOC = 360° - 90° - 90° - 64° = 116°$

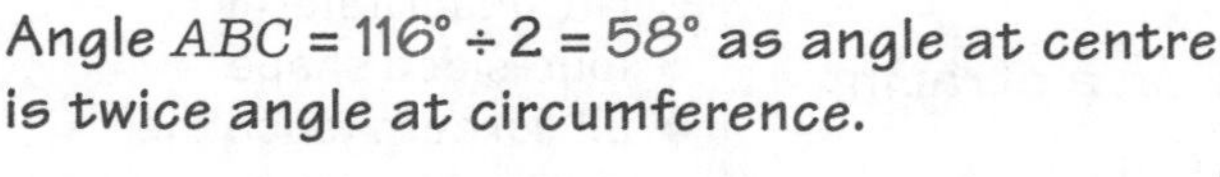

Angle $ABC = 116° \div 2 = 58°$ as angle at centre is twice angle at circumference.

Finding acute angles

A, B and X are points on a circle centre O. YX is a tangent. Angle $ABX = 86°$. Angle $AYX = 52°$
Work out the size of angle OAY.

Solution

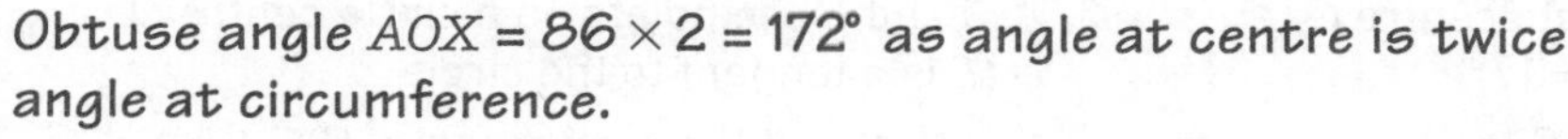

Obtuse angle $AOX = 86 \times 2 = 172°$ as angle at centre is twice angle at circumference.

Angle $AOX = 360° - 172° = 188°$ as angles at a point equal $360°$

Angle $OXY = 90°$ as angle between tangent and radius $= 90°$

Angle $OAY = 360° - 90° - 52° - 188° = 30°$ as sum of angles in quadrilateral is $360°$

Angles in a segment

A, B, C and D are points on a circle centre O.
Angle $ABD = 65°$

a Write down the size of angle ACD.
b Work out the size of angle AOD.

Solution

a Angle $ACD = 65°$ as angles in the same segment are equal.
b Angle $AOC = 65° \times 2 = 130°$ as angle at centre is twice angle at circumference.

Finding a missing length

A, B, C and D are points on a circle.
AXB is a diameter. $BX = 3$ cm, $CX = 9$ cm and $DX = 4$ cm. Find the length of the radius.

Solution

$AX \times XB = CX \times XD$

$AX \times 3 = 9 \times 4$

$AX = (9 \times 4) \div 3 = 12$ cm

$AB = AX + XB = 12 + 3 = 15$ cm

So radius $= 15 \div 2 = 7.5$ cm

Watch out!

Make it clear which angle you have found in each step. So write 'angle $AOC = 116°$' not just '116°'

Exam tip

Use three letter notation for angles; remember the middle letter tells you where the angle is. Label any angles you find on the diagram.

Remember

Correct reasoning and terminology must be used. For instance, use **alternate segment**, not **alternate angle** and use **circumference** not **perimeter**.

Angles in an alternate segment

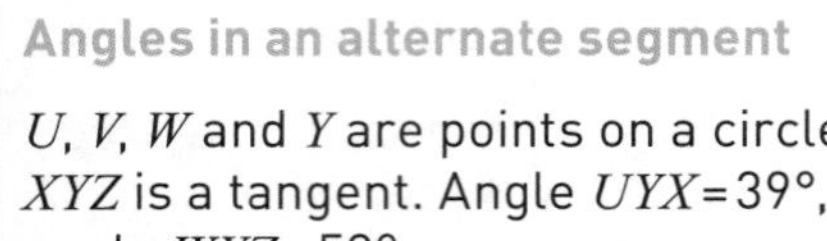

U, V, W and Y are points on a circle. XYZ is a tangent. Angle $UYX = 39°$, angle $WYZ = 52°$

a Write down the size of angle UWY.

b Work out the size of angle UVW.

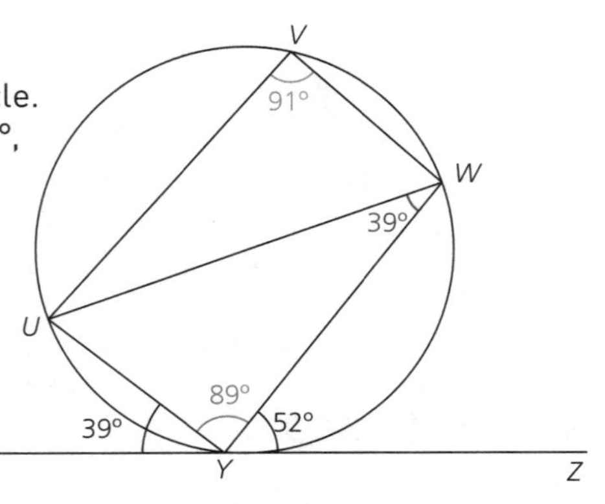

Solution

a 39°, angles in alternate segment equal.

b Angle $UYW = 180° - 39° - 52° = 89°$ as angles on a straight line add to 180°

Angle $UVW = 180° - 89° = 91°$ as opposite angles of a cyclic quadrilateral add to 180°

Remember

Cyclic quadrilateral
- four-sided shape
- all corners touch the circumference
- opposite angles add up to 180°

Exam-style questions

1 A, B, C and D are points on a circle centre O. Angle $DAB = 55°$ and Angle $DBC = 17°$

Work out the size of angle ODC. Give reasons for your answer. [4]

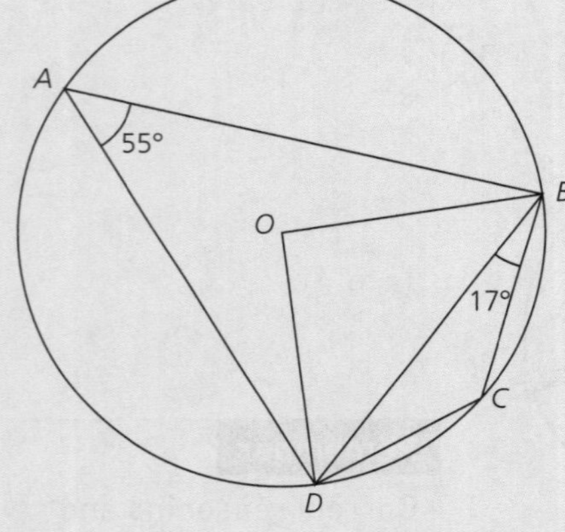

2 A, B and C are points on a circle centre O. DAE is a tangent to the circle.

a Work out the size of angle ABC. Give reasons for your answer. [2]

b Work out the size of angle BAE. Give reasons for your answer. [2]

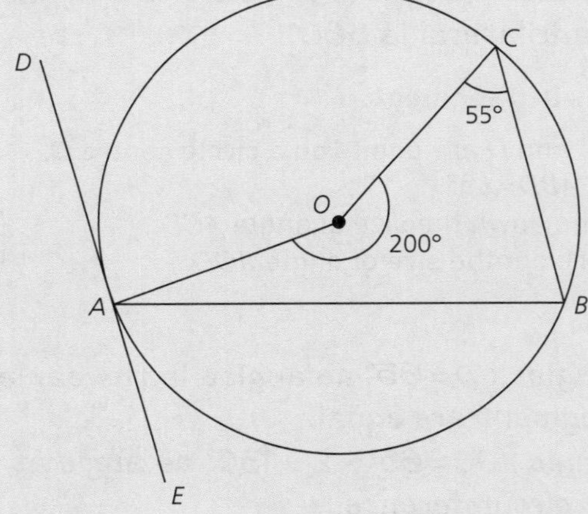

Short answers on page 142

Full worked solutions online

CHECKED ANSWERS ONLINE

Review questions: Geometry and transformations

1 AB and CD are two parallel lines. Angle $AEF = 32°$

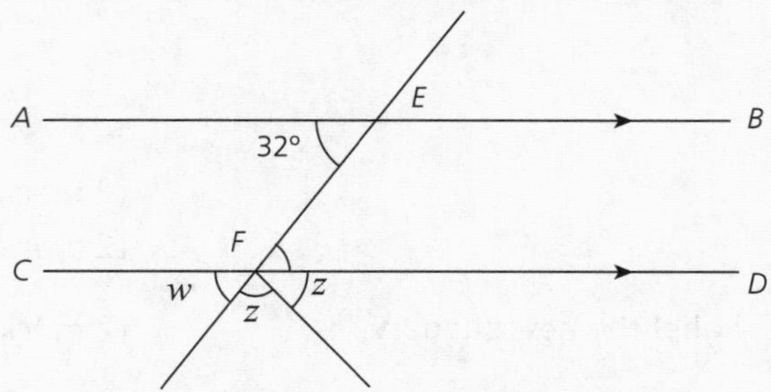

a Write down the size of angle DFE. (1 mark)

b Work out the size of angle BEF. (1 mark)

c Find the size of the angle marked w. (1 mark)

d Work out the value of z. (2 marks)

2 The diagram shows a parallelogram $ABCD$.

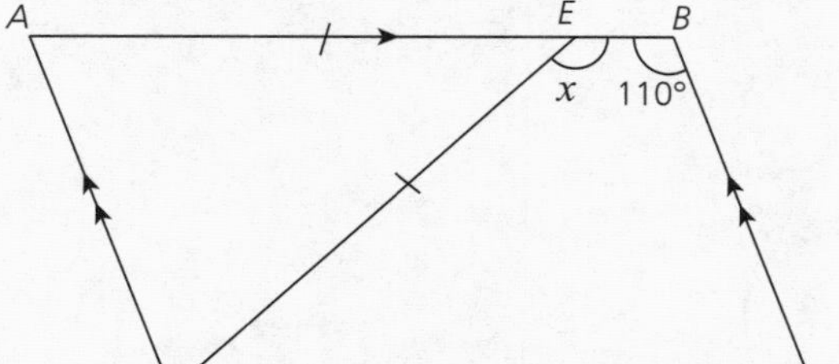

Angle $ABC = 110°$

E is the point on AB such that $AE = DE$.

Angle $DEB = x°$

Work out the value of x. Give reasons for each stage of your working. (4 marks)

3 Sonia leaves point A and walks on a bearing of 046° for 500 m to point B. At point B, Sonia turns and walks on a bearing of 140° for 800 m to point C.
Calculate the bearing of C from A. Give your answer correct to 3 significant figures. (4 marks)

4 B, D and E are points on a circle centre O.

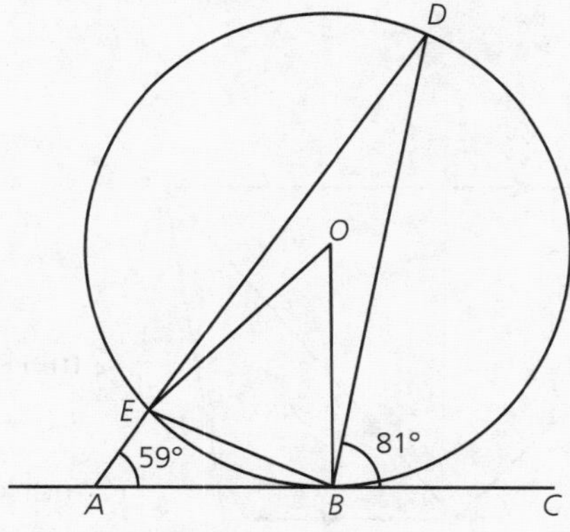

ABC is a tangent.

DEA is a straight line.

Angle $CBD = 81°$

Angle $EAB = 59°$

a Work out the size of angle BED. Explain your answer. (1 mark)

b Work out the size of angle BOE. Explain your answer. (2 marks)

5 ABC is a triangle. Construct

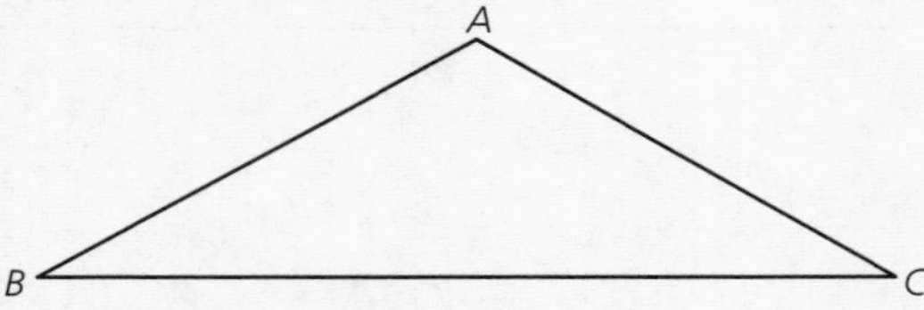

a the bisector of angle ACB (2 marks)

b the perpendicular bisector of the line AB. (2 marks)

6 a On the grid, translate shape **U** by the vector $\begin{pmatrix} -7 \\ -9 \end{pmatrix}$. Label the new shape **V**. (2 marks)

b Describe fully the single transformation that maps shape **U** onto shape **W**. (2 marks)

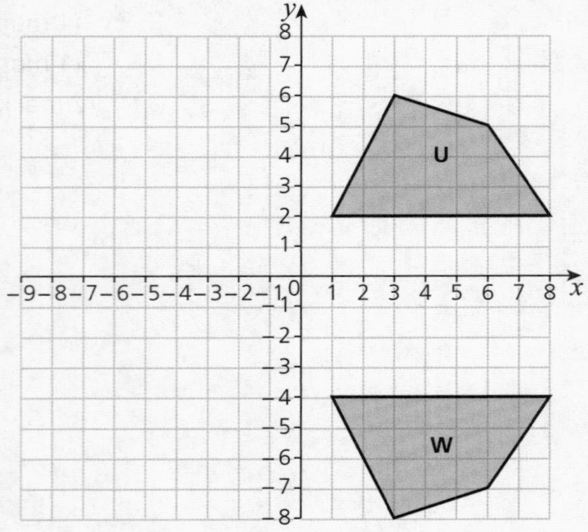

7 Here is the sketch of a curve $y = a\cos bx + c$ Find the values of a, b and c. (3 marks)

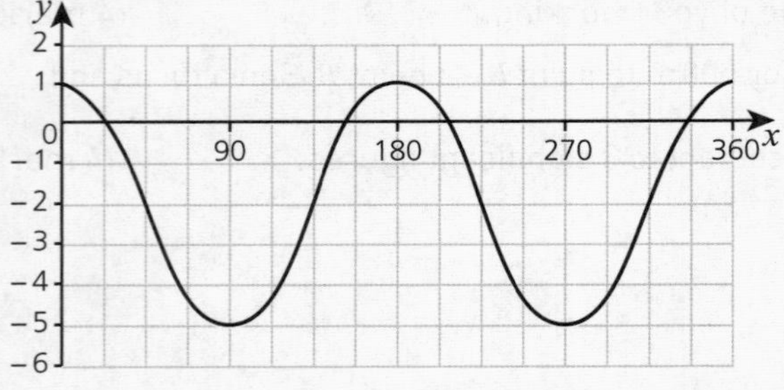

8 The diagram shows a quadrilateral $ABCD$.

$\overrightarrow{CD} = \mathbf{b}$, $\overrightarrow{BC} = \mathbf{a}$, $\overrightarrow{BA} = 2\overrightarrow{CD}$

X is a point on $\overrightarrow{BD}$ such that $BX:XD = 2:1$

a Use a vector method to prove that CXA is a straight line. (2 marks)

Y is on $\overrightarrow{BC}$ such that $\overrightarrow{BY} = \frac{2}{3}\overrightarrow{BC}$.

b Is $\overrightarrow{YX}$ parallel to $\overrightarrow{CD}$? Give reasons for your answer. (2 marks)

Short answers on pages 142–143

Full worked solutions online

CHECKED ANSWERS ONLINE

Target your revision: Statistics and probability

Check how well you know each topic by answering these questions. If you struggle, go to the page number in brackets to revise that topic.

1 Find averages

8 people took a history test. Here are the scores:

21 31 14 17 31 24 31 25

a Find the modal result.

b Find the mean result.

c Find the median result.

(see page 109)

2 Find the range

Here are 11 numbers:

1 3 7 12 4 7 10 11 12 13 9

Find the range.

(see page 109)

3 Solve problems with frequency tables

The distance (d miles) office workers travel to work is shown in the table.

Distance (d miles)	Frequency
$0 < d \leqslant 10$	17
$10 < d \leqslant 20$	12
$20 < d \leqslant 30$	9
$30 < d \leqslant 40$	3

a Write down the modal class.

b Work out an estimate for the mean number of miles travelled.

c Write down the class containing the median distance.

(see page 110)

4 Solve problems involving the mean

A group of students consists of 15 boys and 10 girls.

The mean age for the boys is 19 years.

The mean age for the girls is 16 years.

Find the mean age for the whole group.

(see page 109)

5 Find raw data from the mean

Seven numbers have a mean of 6

Six of the numbers are:

9 4 7 3 8 2

Find the seventh number.

(see page 109)

6 Solve problems using histograms

The table gives information about the time, in minutes, students took to read a popular book.

Time (t minutes)	Frequency
$0 < t \leqslant 30$	45
$30 < t \leqslant 90$	120
$90 < t \leqslant 120$	147
$120 < t \leqslant 300$	90

a Draw a histogram to show the information in the table.

b Estimate the number of people who took over 100 minutes to read the book.

(see page 112)

7 Solve cumulative frequency problems

The grouped frequency table gives information about the lengths of some ropes.

Length (x metres)	Frequency
$0 < x \leqslant 1$	10
$1 < x \leqslant 2$	15
$2 < x \leqslant 3$	26
$3 < x \leqslant 4$	66
$4 < x \leqslant 5$	32

a Draw a cumulative frequency table.

b Draw the cumulative frequency graph.

c Find the median length of the ropes.

d Estimate the interquartile range.

e Estimate the number of ropes longer than 1.5 m.

(see page 114)

8 Use medians and quartiles

Here are the times, in minutes, Jon takes to complete a set of puzzles:

11 15 9 16 17 21 12 13 17 21 19

a Find the median time.

b Find the upper and lower quartile.

c Find the interquartile range.

Peter then does the same puzzles. The interquartile range of Peter's times is 4 minutes.

d Who has the more consistent times?
Give a reason for your answer.

(see page 109)

9 Understand probability

On the probability scale, mark with a cross (x) the probability that:

i picking a day of the week contains the letter *a*,

ii the first letter of a month of the year, chosen at random, begins with *J*,

iii you can score a 7 on a 6-sided dice.

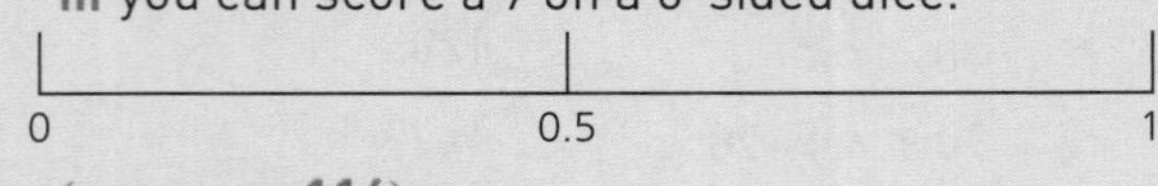

(see page 116)

10 Solve problems with probability

Beth has a spinner. When the spinner is spun the probability of the spinner landing on 3 is $\frac{3}{10}$

Beth spins the spinner 50 times. Find an estimate for the number of times the spinner lands on the number 3

(see page 116)

11 Understand mutually exclusive events

A box has different coloured balls of wool. Each ball of wool is red or white or blue. Sam takes a ball of wool at random from the box. The table shows the probability of red and the probability of white being taken.

Colour	Red	White	Blue
Probability	0.3	0.45	

a Find the probability Sam takes a blue ball of wool.

b Find the probability he takes a red ball of wool or a white ball of wool.

(see page 116)

12 Solve probability problems

Karl is waiting to board his plane. The probability that the plane will leave on time is 0.25

The probability that the plane will be delayed is 0.6

Find the probability that the plane will leave on time or will be delayed.

(see page 116)

13 Understand independent events

A 6-sided dice is thrown twice. Find the probability that:

i it lands on two even numbers,

ii it lands on one even number and one odd number.

(see page 116)

14 Solve harder probability problems

The probability that Ben is late for school on Monday is 0.3

The probability that Ben is late for school on Tuesday is 0.4

Find the probability that:

a Ben is late on both Monday and Tuesday.

b Ben is late on Monday but on time on Tuesday.

c Ben is on time on both Monday and Tuesday.

(see page 116)

15 Use probability tree diagrams

The probability that it will snow on Thursday is 0.6

If it snows on Thursday the probability that it will snow on Friday is 0.55

If it does not snow on Thursday the probability it will snow on Friday is 0.3

a Draw a tree diagram showing this information.

b Find the probability that it will snow on just one of these two days.

(see page 118)

16 Solve conditional probability problems

There are 4 boys and 8 girls at a playgroup. The teacher selects two of the children at random.

a Draw a tree diagram showing this information.

b Find the probability that two girls were selected.

c Find the probability that at least one boy is selected.

(see page 119)

Short answers on pages 143–144

Full worked solutions online

CHECKED ANSWERS ONLINE

Averages and measures of spread

REVISED

Statistics and probability

Key facts

1 The **mode** is the value which occurs **most often**.

Example: The mode of 1, 3, 1, 0, 1, 0, 0, 1 is 1 as it occurs the most.

2 The **median** is the middle value. Numbers are written in size order, smallest first.
An even number of values has two middle values – the median is the middle of these two values.

An odd number of values has one middle value.

Example: 6, 3, 9, 2 in order is 2, 3, 6, 9
There are two middle values so the median is halfway = 4.5
11, 6, 1, 9, 3 in order is 1, 3, 6, 9, 11
The middle value 6 is the median.

3 The **mean** is the sum of all values divided by how many values there are.

Example: The mean of 4, 8, 7, 5, 9, 4, 8, 3 is sum of values (48) ÷ number of values (8) = 6

4 Mean, median and mode are **averages** – they tell you the typical values of a set of data.

5 Range = highest value – lowest value.

Example: The range of 2, 5, 9, 3 is 9 – 2 = 7

6 **Lower quartile, LQ**, divides the bottom half of data into two halves.
Upper quartile, UQ, divides the upper half of data into two halves.
Interquartile range, IQR = UQ – LQ

7 Range and IQR are **measures of spread** – they tell you how spread out or varied the data is.
A low range or IQR means the data is not very spread out.

To compare two data sets: compare their **average** values **and** their **spread**.

Worked examples

Finding a combination mean

There are 70 pears in a box. Their mean weight is 120 g. 10 pears with a mean weight of 110 g are added to the box. Work out the mean weight of all 80 pears.

Solution

Total weight = (70 × 120) + (10 × 110) = 8400 g + 1100 g = 9500 g

Mean weight = 9500 ÷ 80 = 118.75 g

Finding the interquartile range

Find the interquartile range of 11 33 8 4 10 3 10

Solution

In order: 3 4 8 10 10 11 33
The median is the 4th term i.e. 10

LQ = 4 and UQ = 11 so the IQR = 11 – 4 = 7

Remember

You must order the list first. Do not use the original list.

Watch out!

First find the combined total weight of all pears.

Exam tip

Do not round your answer to the mean to a whole number.

Watch out!

Use the median and IQR when there are one or two values that are much higher or lower than the rest (like 33 in this example) as the median is not affected by extreme values.

Remember

Median is $\left(\frac{n+1}{2}\right)$th term.

Exam-style questions

1 Jon, Sue and Max are three cousins.
The mean age of Jon, Sue and Max is 23 years.
The mean age, in years, of Jon and Sue is 21
Work out Max's age. [3]

2 Here is a list of numbers in order of size:
2, 5, 8, x, y, z.
The numbers have a median of 10, a mean of 9 and a range of 13
Find the value of x, the value of y and the value of z. [4]

3 Here are the marks Tim scored in 11 physics tests: 17 13 20 19 18 14 14 21 12 20 18
Find the interquartile range. [3]

Short answers on page 144
Full worked solutions online

CHECKED ANSWERS ONLINE

Frequency tables

REVISED

Key facts

1 Large amounts of data can be put together in a **frequency table**.
2 A frequency table tells you how many times each value occurs in the data set.
3 From a frequency table you can find the **mode**, **median**, **mean** and **range**.

Worked examples

Solving problems with frequency tables

Rebecca has a biased 6-sided dice. She throws the dice 35 times and records the score of each throw. The table shows the information about her scores.

a Write down the modal score.
b Find the range of scores thrown.
c Find the median score.
d Work out her mean score.

Score	Frequency
1	10
2	9
3	6
4	3
5	2
6	5

Solution

a The modal score is the number with the largest frequency (10). So the modal score is 1

b The range = largest value − smallest value = $6-1=5$

c Use $\frac{n+1}{2}=\frac{35+1}{2}=18$, so the median is the 18th value which is 2

d

Score (x)	Frequency (f)	$x \times f$
1	10	$1\times10=10$
2	9	$2\times9=18$
3	6	$3\times6=18$
4	3	$4\times3=12$
5	2	$5\times2=10$
6	5	$6\times5=30$
Total	35	98

Mean $=\frac{98}{35}=2.8$

Watch out!

Make sure that you do not give a frequency as an answer for the mode or median!

Exam tip

Do not just add up the frequencies and divide by 6. First multiply x and f together and then sum the last two columns.

$$\text{Mean} = \frac{\text{Total of } x \times f}{\text{Total frequency, } f}$$

Exam-style questions

1 The table gives information about the number of goals scored in 60 hockey matches.

Goals scored	0	1	2	3	4	5
Frequency	3	11	25	3	16	2

a Work out the mean number of goals scored. [3]
b Find the median number of goals scored. [2]
c What fraction of the matches had 3 or more goals scored? [2]

2 Tom has a 5 sided spinner. He spins the spinner 75 times and records each score. The table shows information about his scores.

Score	1	2	3	4	5
Frequency	10	21	6	19	19

a Find his median score. [2]
b Work out his mean score. [3]
c Find his modal score. [1]
d Find the range of scores thrown. [1]

Short answers on page 144

Full worked solutions online

CHECKED ANSWERS ONLINE

Grouped frequency tables

REVISED

Key facts

1. Grouped frequency tables group data together in **classes** using **continuous** or **discrete** data.
2. Actual data values are not known so only estimates for averages can be found.
3. To find an estimate of the mean
   a. add a 3rd column to find the **mid-interval** of each class
   b. add a 4th column to show 'frequency x mid-interval'.

Worked examples

Solving problems with grouped frequency tables

This table shows information about the weights, in kilograms, of 70 children.

a Write down the modal class

b Write down the class containing the median weight.

c Find an estimate for the mean weight.

Weight (w, kg)	Frequency
$30 < w \leqslant 40$	8
$40 < w \leqslant 50$	26
$50 < w \leqslant 60$	23
$60 < w \leqslant 70$	13

Solution

a The modal class is the one with the highest frequency (26). So the modal class is $40 < w \leqslant 50$.

b Use $\frac{n+1}{2} = \frac{70+1}{2} = 35.5$ i.e. the median is between the 35th and 36th values. These are both in the 3rd class, so $50 < w \leqslant 60$ contains the median.

c Add extra columns for 'mid-interval' and 'frequency x mid-interval'.

Weight (w, kg)	Frequency (f)	Mid-interval (x)	fx
$30 < w \leqslant 40$	8	35	$8 \times 35 = 280$
$40 < w \leqslant 50$	26	45	$26 \times 45 = 1170$
$50 < w \leqslant 60$	23	55	$23 \times 55 = 1265$
$60 < w \leqslant 70$	13	65	$13 \times 65 = 845$
Total	70		3560

Mean $= 3560 \div 70 = 50.857 = 50.9$ kg (to 3 s.f.)

Remember

The first 8 values are in the 1st class.
The first 8 + 26 = 34 values are in class 1 or 2
So the 35th and 36th values are in the 3rd class.

Remember

Mid-intervals are found by adding the end values of the class and dividing by 2, i.e. $\frac{30+40}{2} = 35$

Watch out!

Do not add up the mid-interval column.

Exam tip

Check your answer looks sensible.
Do not find the sum of the frequencies and divide by the number of classes.

Exam-style question

The table shows information about the length, in cm, of 75 spiders.

Length (L cm)	Frequency
$0 < L \leqslant 1$	7
$1 < L \leqslant 2$	9
$2 < L \leqslant 3$	21
$3 < L \leqslant 4$	27
$4 < L \leqslant 5$	11

a Write down the class containing the median length. [2]

b Estimate the range of the lengths. [1]

c Find an estimate for the mean length, giving your answer to 1 d.p. [4]

Short answers on page 144
Full worked solutions online

CHECKED ANSWERS ONLINE

Histograms

REVISED

Key facts

1. Histograms are bar charts where the bars may have unequal widths.
2. The vertical axis is always labelled **frequency density** and there are **no gaps** between bars.
3. Frequency density $= \dfrac{\text{frequency}}{\text{class width}}$
4. Rearranging the formula for frequency density gives:
   Frequency = frequency density × class width
   So the area of each bar gives you the frequency.

Worked examples

Drawing a histogram

The table below gives information about the distance, in miles, travelled by 240 people to get to a concert. Draw a histogram to show this information.

Solution

Distance (D miles)	Freq.	Class width	Freq. density
$0 < D \leqslant 10$	66	10	$66 \div 10 = 6.6$
$10 < D \leqslant 25$	78	15	$78 \div 15 = 5.2$
$25 < D \leqslant 45$	72	20	$72 \div 20 = 3.6$
$45 < D \leqslant 60$	24	15	$24 \div 15 = 1.6$

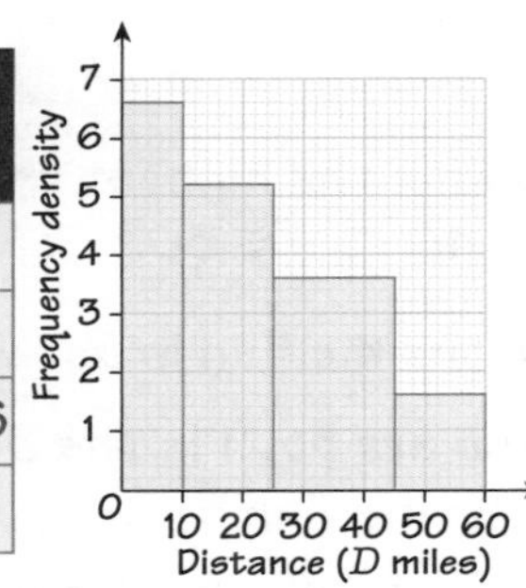

Finding estimates from a histogram

The histogram shows information about the time, in minutes, commuters were delayed on a motorway one day. No one was delayed for over 100 minutes.

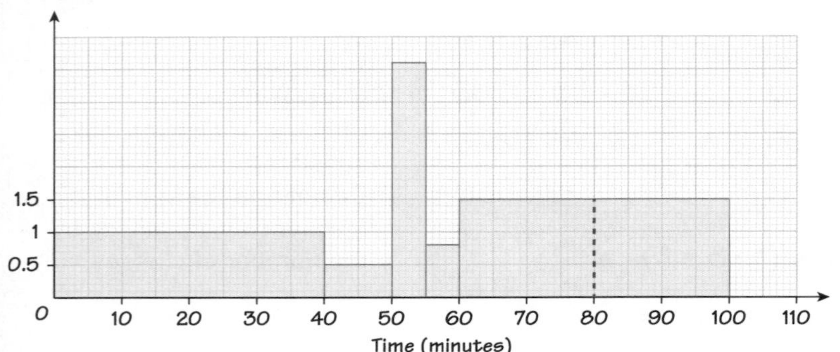

5 passengers experienced delays between 40 and 50 minutes. Estimate the number of commuters who were delayed for more than 80 minutes.

Solution

2nd bar: frequency density $= \dfrac{5}{10} = 0.5$, so every 5 small squares on the vertical axis has a frequency density of 0.5
Frequency $= 1.5 \times 20 = 30$, so 30 commuters were delayed for over 80 minutes.

Remember

To find the class width, for each class, subtract the lower class boundary from the upper class boundary
e.g. 10−0=10
Frequency density is plotted on the vertical axis.

Exam tip

Add a class width column and a frequency density column to the table.

Exam tip

Use a sensible scale on the grid to keep accuracy.
Do not have a histogram that is too small – you'll be given a grid in the exam, so use as much of it as you can!

Watch out!

There are no values on the vertical scale.
Use the information in the question to work out the frequency density scale.
Remember: frequency density $= \dfrac{\text{frequency}}{\text{class width}}$
Once you have found the frequency density of the 2nd bar you can use it to mark up the vertical axis.

Exam tip

Correctly labelling the frequency density axis can get you marks!

Completing a table and a histogram

The incomplete histogram and table gives information about the flight times to various holiday destinations.

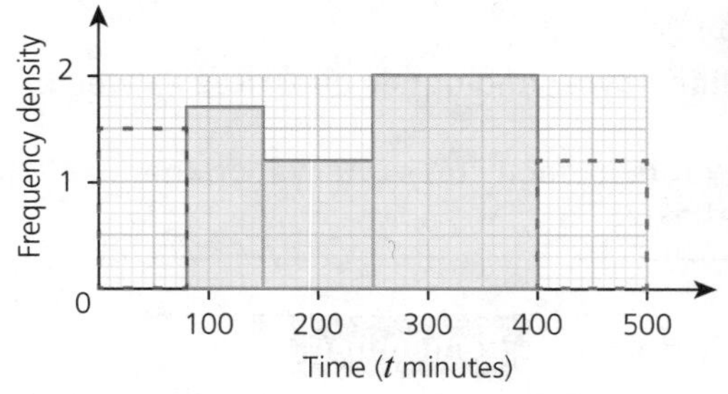

Time (t minutes)	Frequency
$0 < t \leqslant 80$	120
$80 < t \leqslant 150$	119
$150 < t \leqslant 250$	
$250 < t \leqslant 400$	
$400 < t \leqslant 500$	130

a Use the histogram to complete the table.
b Use the table to complete the histogram.

Solution

a Using class 2: Frequency density $= \frac{119}{70} = 1.7$ and the bar height is 17 squares. So each small square has a frequency density of $1.7 \div 17 = 0.1$

Bar for the 3rd class is 12 small squares high so frequency is $1.2 \times 100 = 120$

Bar for the 4th class is 20 small squares high so frequency is $2 \times 150 = 300$

b For the 1st class: Frequency density $= \frac{120}{80} = 1.5$ so draw a bar 15 small squares high.

For the 5th class: Frequency density $= \frac{130}{100} = 1.3$ so draw a bar 13 small squares high between 400 and 500 minutes.

Exam-style questions

1 The histogram shows information about the time, t minutes, patients spent waiting at a busy hospital department for their appointment one day.

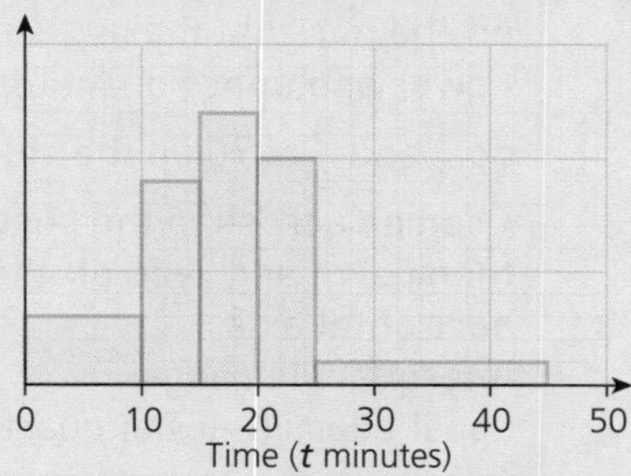

a Find the percentage of patients who waited over 25 minutes. [3]

b 60 patients waited between 15 and 20 minutes. Find the number of patients who waited less than 5 minutes that day. [2]

2 Cat grows carrots in her garden. The table shows information about the weight, in grams, of some of her carrots.

Weight of carrots (w grams)	Frequency
$0 < w \leqslant 30$	57
$30 < w \leqslant 40$	15
$40 < w \leqslant 55$	18
$55 < w \leqslant 74$	19
$74 < w \leqslant 95$	42

a Draw a histogram to show the above information. [3]

b What percentage of carrots weigh less than 55 grams? [2]

c In which interval does the median lie? [2]

d Find an estimate for the number of carrots that weigh 50 grams or less. [2]

Short answers on page 144

Full worked solutions online

CHECKED ANSWERS ONLINE

Statistics and probability

Cumulative frequency

REVISED

Key facts

1. **Cumulative frequency** means to add up the frequency as you go along.
2. A cumulative frequency curve can be drawn to find the **median**, **lower** and **upper quartiles** as well as the **interquartile range**.
3. Cumulative frequency curves are plotted using the **highest value** in each class interval on the horizontal axis and the **cumulative frequency** on the vertical axis.

Worked examples

Drawing a cumulative frequency curve and estimating averages

The table gives information about the heights of 70 teachers at a school.

a Draw a cumulative frequency table.
b Draw a cumulative frequency graph.
c Use your graph to estimate
  i the median
  ii the lower quartile
  iii the upper quartile.
d Estimate the interquartile range.
e Estimate the number of teachers taller than 170 cm.

Height (h cm)	Frequency
$150 < h \leqslant 160$	7
$160 < h \leqslant 170$	20
$170 < h \leqslant 180$	11
$180 < h \leqslant 190$	17
$190 < h \leqslant 200$	15

Solution

a

Height (h cm)	Frequency	Cumulative frequency
$150 < h \leqslant 160$	7	7
$160 < h \leqslant 170$	20	$7 + 20 = 27$
$170 < h \leqslant 180$	11	$27 + 11 = 38$
$180 < h \leqslant 190$	17	$38 + 17 = 55$
$190 < h \leqslant 200$	15	$55 + 15 = 70$

b

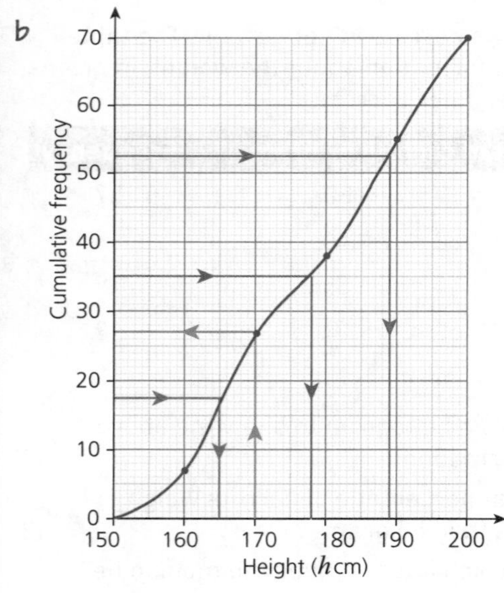

c i Median = 178 cm
  ii Lower quartile = 165 cm
  iii Upper quartile = 189 cm

d Interquartile range = 189 − 165 = 24

e From the graph the number of teachers who are 170 cm or under = 27
So the number of teachers who are taller than 170 cm is 70 − 27 = 43

Remember

Any values from the graph will be estimates because they are based on group data and actual data values are not known.

Remember

The last value in the cumulative frequency column must equal the sum of the frequency column.

Remember

Median: go half of 70 up the side. Then go across to the curve then down and read off the horizontal axis.
Lower and upper quartiles: go $\frac{1}{4}$ and $\frac{3}{4}$ of 70 up the side. Then go across to the curve then down and read off the horizontal axis.
Interquartile range = upper quartile − lower quartile.

Working with a cumulative frequency curve

A survey was carried out to find the duration, in minutes, of international telephone calls. The results are shown in the cumulative frequency graph.

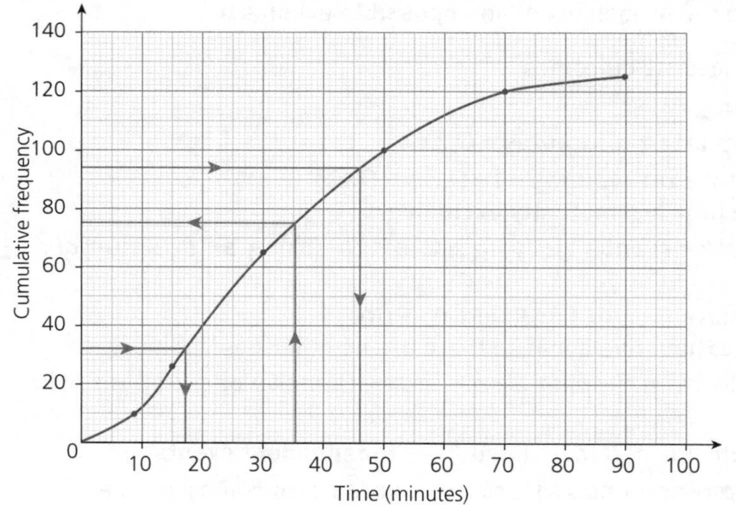

a What was the total number of calls in the survey?
b Find an estimate for the number of calls over 35 minutes.
c Find an estimate for the interquartile range of the lengths of calls.

Solution

a From the graph the highest value vertically, of the curve, is 125 so this is the number of calls in the survey.

b From the graph the number of calls 35 minutes and under = 75
So the number of calls over 35 minutes is $125 - 75 = 50$

c Lower quartile = 17 and Upper quartile = 46
Interquartile range = $46 - 17 = 29$

Remember

Cumulative frequency gives a running total of the frequencies.

Exam tip

Make sure you read the scale correctly.

The quartiles are not $\frac{1}{4}$ or $\frac{3}{4}$ up the side. Do lines from $125 \div 4$ and $3 \times (125 \div 4)$ on the vertical axis. Go across to the curve then down and read off the horizontal axis.

Exam-style question

The table shows information about the weights, in grams, of 85 letters.

Weight (w grams)	Frequency
$0 < w \leqslant 5$	4
$5 < w \leqslant 10$	28
$10 < w \leqslant 15$	30
$15 < w \leqslant 20$	19
$20 < w \leqslant 25$	4

a Draw a cumulative frequency graph to show the information in the table. [2]

b Find an estimate for the percentage of letters that weigh less than 20.5 g. [2]

Short answers on page 144
Full worked solutions online

Probability

REVISED

Key facts

1 The probability of a **certain** event is **1** and the probability of an **impossible** event is **0**

2 Probability of an event = $\frac{\text{number of outcomes}}{\text{total number of possible outcomes}}$

3 P(event happening) + P(event not happening) = 1

Example: The probability of getting up late on Monday = 0.35

So the probability of not getting up late on Monday = 1 − 0.35 = 0.65

4 Events which cannot happen at the same time are **mutually exclusive**.

Example: When you throw a fair dice, the events 'getting a 2' and 'getting an odd number' are mutually exclusive.

5 If two results A and B are **mutually exclusive** then **P(A or B) = P(A) + P(B)**.

6 Outcomes that cover all possibilities are **exhaustive** because there are no other possibilities.

Example: The results of heads or tails from flipping a coin are exhaustive because no other results are possible.

7 If the outcome of A does not affect the outcome of B then A and B are **independent events**.

Example: The result from flipping a coin does not affect the score when rolling a dice.

8 If events A and B are **independent** then P(A and B) = P(A) × P(B).

Worked examples

Finding probabilities

Rose has a piece of fruit each morning for breakfast. She chooses at random an apple, pear or banana. The probability she chooses an apple is 0.3 and the probability she chooses a pear is 0.15

a Work out the probability that she chooses a banana.

b There are 30 days in June. Find an estimate for the number of days in June in which she chooses an apple.

Solution

a P(banana) = 1 − (0.3 + 0.15) = 1 − 0.45 = 0.55

b Estimated number of days when Rose chooses an apple = 30 × 0.3 = 9

Working with mutually exclusive probabilities

Jon is waiting to catch a train. The probability that his train is on time is 0.18 and the probability that the train is late is 0.5

Find the probability that the train will be either on time or late.

Solution

P (on time or late) = P(on time) + P(late) = 0.18 + 0.5 = 0.68

Working with independent events

Mark travels to school either by bus or by bike. On any given day, the probability he will travel by bike is 0.6 and the probability of being late for school is 0.3. When he travels by bus, the probability he will be late for school is 0.2

a Work out the probability of Mark travelling by bus and being late for school.

b Work out the probability that Mark will not be late for school.

Solution

a P(bus and late) = P(bus) × P(late) = (1 − 0.6) × 0.2 = 0.4 × 0.2 = 0.08

b Mark has two options: Go by bike and not be late or go by bus and not be late.

Remember

The sum of all probabilities is equal to 1

Exam tip

It is important to recognise events that are mutually exclusive. Jon's train can't be both late and on time, so these events are mutually exclusive.
Remember to add the probabilities.

Exam tip

Do not just multiply the probabilities of not being late together.

P(bike and not late) = $0.6 \times 0.7 = 0.42$
P(bus and not late) = $0.4 \times 0.8 = 0.32$
Therefore P(Mark not late) = $0.42 + 0.32 = 0.74$

Finding probabilities from Venn diagrams

Peter asked 75 people which sport they enjoyed from football, rugby and swimming. The Venn diagram shows this information.

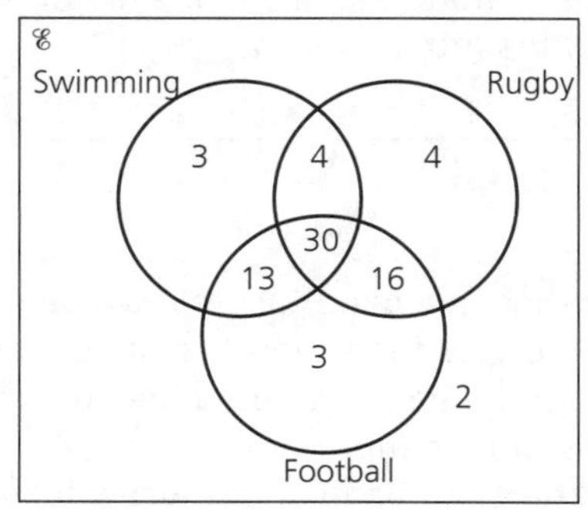

Find the probability that a person:

a i likes football only
ii likes exactly one sport only
iii likes swimming and football but not rugby

b Find P(swimming ∩ rugby ∩ football)

Solution

a i $P(\text{football only}) = \frac{3}{75}$

ii $P(\text{one sport only}) = \frac{3+4+3}{75} = \frac{10}{75}$

iii $P(\text{swimming and football but not rugby}) = \frac{13}{75}$

b $P(\text{swimming} \cap \text{rugby} \cap \text{football}) = \frac{30}{75}$

Remember

Remember these links: 'or' with 'add'; 'and' with 'multiply'.

Exam tip

You do not need to cancel your probability answer if it is a fraction.

Remember

∩ means intersection.

Exam-style questions

1 A four-sided spinner is spun. It can land on pink or yellow or green or blue.
The table shows the probabilities of it landing on a particular colour.

Colour	Pink	Yellow	Green	Blue
Probability	0.17	0.22	0.35	

a Find the probability that the spinner lands on blue. [2]

b The spinner is spun 1000 times. Find an estimate for the number of times the spinner lands on yellow. [2]

2 A bag has 20 white counters and 15 black counters. A counter is picked at random and replaced.

a Work out the probability that both counters are black. [2]

b Work out the probability that one counter is black and the other counter is white. [2]

c Work out the probability of picking a black counter or a white counter. [1]

3 Tom has to catch two trains. The probability that the first train will be late is 0.35 and the probability that the second train will be late is 0.2

a Find the probability that both trains will be late. [2]

b Find the probability that neither train will be late. [2]

4 The table shows how a group of students travel to school.

	Bike	Walk	Bus	Total
Boys	40	55	80	175
Girls	37	67	41	145
Total	77	122	121	320

a Find the probability that a student chosen at random who walks to school is a boy. [2]

b Find the probability that a girl chosen at random catches a bus. [2]

c Find the probability that a student of the school chosen at random is a boy who walks to school. [2]

Short answers on page 144
Full worked solutions online

CHECKED ANSWERS ONLINE

Probability tree diagrams

REVISED

Key facts

1. **Tree diagrams** are a useful way to calculate probabilities when there are combined events.
2. To calculate probabilities **multiply** along the branches to get the end probability.
3. The sum of all the end probabilities must equal 1.
4. P(at least 1) = 1 – P(none)

Worked examples

There are 5 red beads and 7 blue beads in a bag. Karl takes a bead at random from the bag, notes the colour. He then replaces the counter. Andrew takes a bead at random from the bag and notes the colour.

a Draw a tree diagram to show this information.

b Find the probability that Karl takes a red bead and Andrew takes a blue bead.

c Work out the probability that both beads taken are the same colour.

Solution

a

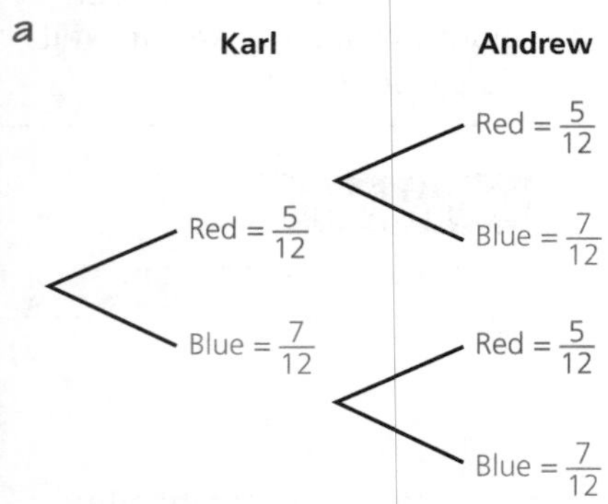

b P(Karl Red and Andrew Blue) = P(Karl Red) × P(Andrew Blue)

$$= \frac{5}{12} \times \frac{7}{12} = \frac{35}{144}$$

c P(R and R or B and B)

$$= P(R \text{ and } R) + (B \text{ and } B)$$

$$= \left(\frac{5}{12} \times \frac{5}{12}\right) + \left(\frac{7}{12} \times \frac{7}{12}\right) = \left(\frac{25}{12}\right) + \left(\frac{49}{144}\right) = \frac{74}{144}$$

Watch out!

There are two possible outcomes for Andrew so two branches need to be drawn. Karl **replaces** the counter.

Exam tip

What Karl picks and what Andrew picks are **independent** so multiply the probabilities.

Exam tip

Both beads being the same colour means it is mutually exclusive, so add the probabilities.

Exam-style questions

1 Angela goes to school either by car or by bus. On any day the probability that she goes to school by car is $\frac{1}{3}$. When she goes by car the probability she is late is $\frac{1}{4}$. When she goes by bus the probability she is late is $\frac{5}{6}$.

a Draw a probability tree diagram to show this information. [2]

b Work out the probability that on a day Angela goes to school, she goes by car and is late. [2]

c Work out the probability that on a day Angela goes to school, she is not late. [2]

2 David takes part in two 400 m relay races. The probability he wins the first race is 0.65 The probability he wins the second race is 0.55

a Draw a probability tree diagram to show this information. [2]

b Work out the probability that David will win both races. [2]

David takes part in a third race. The probability he wins the third race is 0.35

c Work out the probability that he wins

(i) exactly one of the three races [3]

(ii) at least one of the three races. [3]

Short answers on pages 144–145
Full worked solutions online

CHECKED ANSWERS ONLINE

Conditional probability

REVISED

Key facts

1. Conditional probability describes the situation where the probability of an event depends on the outcome of another event.
2. Conditional probability questions often contain **without replacement** within the question.

Worked examples

Solving problems with conditional probability

A bag contains 7 buttons, 3 are red and 4 are blue. A button is removed and not replaced. Another button is then removed. If the first button removed is red, find the probability that:

a the second button will be red b both buttons are blue.

Solution

a $P(\text{second button red}) = \frac{2}{6}$ b $P(\text{both blue}) = \frac{4}{7} \times \frac{3}{6} = \frac{12}{42}$

Conditional probability and tree diagrams

A bag has 6 red crayons and 4 green crayons. Sadhiv takes two counters from the bag at random, without replacement.

a Work out the probability that he takes two red crayons.
b Work out the probability that he takes both crayons of the same colour.
c Work out the probability that he takes one of each colour.
d Work out the probability that he takes at least one green crayon.

Solution

1st pick: $R = \frac{6}{10}$, $G = \frac{4}{10}$

2nd pick: after R: $R = \frac{5}{9}$, $G = \frac{4}{9}$; after G: $R = \frac{6}{9}$, $G = \frac{3}{9}$

a $P(R \text{ and } R) = \frac{6}{10} \times \frac{5}{9} = \frac{30}{90}$

b $P(R \text{ and } R) \text{ or } P(G \text{ and } G)$

$= \frac{6}{10} \times \frac{5}{9} + \frac{4}{10} \times \frac{3}{9}$

$= \frac{30}{90} + \frac{12}{90} = \frac{42}{90}$

c $P(R \text{ and } G) \text{ or } P(G \text{ and } R)$

$= \left(\frac{6}{10} \times \frac{4}{9}\right) + \left(\frac{4}{10} \times \frac{6}{9}\right) = \frac{48}{90}$

d $P(\text{At least 1 } G) = 1 - P(R \text{ and } R) = 1 - \frac{30}{90} = \frac{60}{90}$

Remember

If a red button is removed first, 2 red buttons are left with 6 buttons in total.
If a blue button is removed first, 3 blue buttons are left and 6 buttons in total.

Exam tip

Draw a tree diagram to determine which numbers are to be multiplied together.

Watch out!

This is a 'without replacement' question. Both numerator and denominator change in the 2nd pick.

Exam tip

Do not cancel your fractions at any point.
Remember:
P(at least one) = 1 – P(none)

Exam-style questions

1 There are 20 coloured sweets on a tray. 14 of the sweets are green and 6 are purple. Zoe takes at random one of the sweets and eats it. She then takes at random another sweet and eats it.

a Draw a tree diagram to show this information. [3]
b Work out the probability that Zoe eats one green sweet and one purple sweet. [3]
c Work out the probability that Zoe eats at least one purple sweet. [3]

2 At a party there are 5 ham sandwiches, 6 cheese sandwiches and 4 egg sandwiches. Emily takes at random a sandwich and then Matthew takes at random a sandwich.
Work out the probability that Emily and Matthew both take a sandwich of the same type. [3]

3 There are 7 strawberry flavoured ice lollies and 5 orange flavoured ice lollies in a box. Arti takes at random an ice lolly from the box and eats it. Then Barak takes at random an ice lolly from the box and eats it. Work out the probability that there are exactly 4 orange ice lollies left. [3]

Short answers on page 145
Full worked solutions online

CHECKED ANSWERS ONLINE

Review questions: Statistics and probability

1 Here is a list of six numbers written in order of size:

$4 \quad 7 \quad 9 \quad x \quad y \quad z$

The numbers have a median of 13 and a mean of 15

y is 3 more than x.

Find the numbers x, y and z (4 marks)

2 There are 70 tomatoes in a box. Their mean weight is 42 g.

There are 20 tomatoes in another box. Their mean weight is 38 g.

Find the mean weight of the 90 tomatoes. (3 marks)

3 The table shows the amount of Euros tourists spent in a shop on one particular day.

Money spent (€ Euros)	Frequency
$0 < € \leqslant 15$	12
$15 < € \leqslant 30$	13
$30 < € \leqslant 45$	22
$45 < € \leqslant 60$	45
$60 < € \leqslant 75$	39
$75 < € \leqslant 90$	12

a Find an estimate for the total amount of money spent in the shop on that day. (2 marks)

b Find an estimate for the mean amount of money spent in the shop on that day. (2 marks)

4 The table gives information about the time taken for some athletes to finish a race.

Time (t minutes)	Frequency
$0 < t \leqslant 50$	10
$50 < t \leqslant 75$	25
$75 < t \leqslant 150$	30
$150 < t \leqslant 160$	20
$160 < t \leqslant 200$	15

a Use the information in the table to draw a histogram. (3 marks)

b Estimate the number of athletes who took over 90 minutes to complete the race. (3 marks)

5 The table gives information about the times taken by some people to complete a race.

Time (t minutes)	Frequency
$0 < t \leqslant 15$	6
$15 < t \leqslant 25$	8
$25 < t \leqslant 30$	21
$30 < t \leqslant 45$	9

a Use the information in the table to complete the histogram. (3 marks)

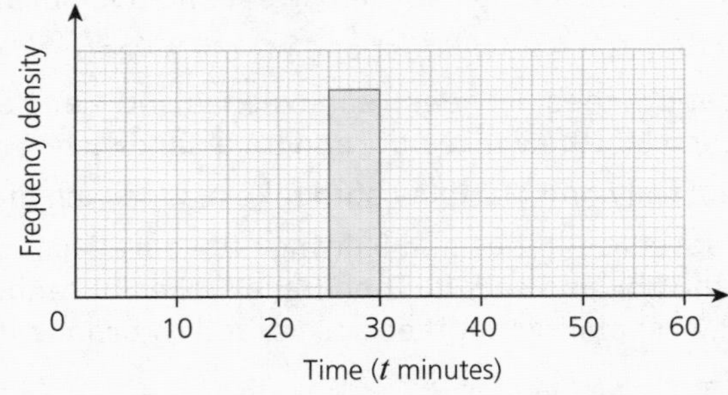

b Find an estimate for the number of people who took between 20 and 40 minutes to complete the race. (3 marks)

6 The cumulative frequency table shows information about the time, in minutes, trains were delayed one morning.

Time (t minutes)	Cumulative frequency
$0 < t \leqslant 10$	5
$0 < t \leqslant 20$	15
$0 < t \leqslant 30$	33
$0 < t \leqslant 40$	66
$0 < t \leqslant 50$	87
$0 < t \leqslant 60$	102
$0 < t \leqslant 70$	113
$0 < t \leqslant 80$	126
$0 < t \leqslant 90$	135
$0 < t \leqslant 100$	140

a Draw a cumulative frequency graph for the table. (2 marks)

b Use your graph to estimate the median. (2 marks)

c Use your graph to estimate the interquartile range. (2 marks)

d Use your graph to find an estimate for the number of trains delayed by more than $1\frac{1}{4}$ hours. (2 marks)

7 Peter is practising his penalty shootouts with a ball.
When Peter kicks the ball he can score, the ball can be saved or the ball goes over the net.
He takes two penalties.
The probability he kicks the ball over the net is 0.2
The probability he scores is 4 times the probability that the ball is saved.

a Draw a tree diagram to show this information. (4 marks)

b Find the probability that Peter scores at least one goal. (3 marks)

8 Ofer puts 15 apples and 10 pears in a bowl.
Linda takes at random a piece of fruit from the bowl and eats it.
She then takes another piece of fruit and eats it.
Find the probability that she eats two apples. (3 marks)

Short answers on page 145

Full worked solutions online

CHECKED ANSWERS ONLINE

Revision tips

Before your exam...

REVISED

- *Start revising early* – half an hour a day for 6 months is better than cramming in all-nighters in the week before the exam. Little and often is the key.
- *Don't procrastinate* – you won't feel more like revising tomorrow than you do today!
- Put your phone on *silent* while you revise – don't get distracted by a constant stream of messages from your friends.
- Make sure your *notes are in order* and nothing is missing.
- *Be productive* – don't waste time colouring endless revision timetables. Make sure your study time is actually spent revising!
- Use the '*Target your revision*' sections to *focus your revision* on the topics you find tricky – remember you won't improve if you only answer the questions you could do anyway!
- Don't just read about a topic. *Maths is an active subject* – you improve by answering questions and actually *doing* maths not just reading about it.
- Cover up the solution to an example and then try and answer it yourself.
- Answer as many past exam questions as you can. Work through the '*Review questions*' first and then move on to past papers.
- Try *teaching a friend* a topic. Even *teaching* your teddy bear will do! Teaching something is the best way to learn it yourself – that's why your teachers know so much!
- *Learn all of the angle facts* (pages 86–87) and *circle theorems* (pages 102–104) – make sure you know the correct terminology. You won't score marks for '*Z-angles*' or 'bow-tie theorem', you need to use the correct terms: '*corresponding angles*' or '*angles in the same segment are equal*'. Marks for stating these theorems are often lost in the exam - get a head start on the other candidates by learning them!

Make sure you know these formulae for your exam

Topic	Formula
Number	
Percentages (see page 8)	**Percentages** Percentage increase (or decrease) $= \frac{\text{Difference}}{\text{Original}} \times 100\%$
Powers and roots (see page 10)	**Indices** ● $a^n = \underbrace{a \times a \times a \times \ldots \times a}_{n \text{ factors of } a}$ ● $a^{-n} = \frac{1}{a^n}$ ● $a^0 = 1$ and $a^1 = a$ **Laws of indices** ● $x^m \times x^n = x^{m+n}$ ● $x^m \div x^n = x^{m-n}$ ● $(x^m)^n = x^{mn}$ **Roots** ● $\sqrt{x} = x^{\frac{1}{2}}$ ● $\sqrt[n]{x} = x^{\frac{1}{n}}$ ● $\sqrt[n]{x^m} = x^{\frac{m}{n}}$
Standard form (see page 12)	**Standard form** A number is in standard form when it is written in the form $a \times 10^n$ where n is an **integer** (whole number) and $1 \leqslant a < 10$
Sets (see page 14)	**Set Notation** ● $\in$ means 'is an element of' ● $\notin$ means 'is **not** an element of' ● $n\{A\}$ means the **number of elements** in set A ● $\varnothing$ means the empty set ● $\mathscr{E}$ is the universal set ● $A \cap B$ is the **intersection** of the sets A and B (elements belong to A and B) ● $A \cup B$ is the **union** of the sets A and B (elements belong to A or B or both) ● A' is the complement of A (means not A) ● $C \subset D$ means C is a subset of D
Algebra	
Factorising quadratic expressions (see pages 27–28)	● The **difference of two squares** $x^2 - a^2 = (x+a)(x-a)$ ● A **perfect square** $x^2 + 2ax + a^2 = (x+a)^2$ $x^2 - 2ax + a^2 = (x-a)^2$
Sequences and series (see page 37)	**Arithmetic sequences** The nth term is $a + (n-1)d$ a = 1st term of sequence d = common difference

Functions and graphs	
Coordinates (see page 44)	**Coordinates** For two points (x_1, y_1) and (x_2, y_2): Midpoint $= \left(\frac{x_1 + x_2}{2}, \frac{y_1 + y_2}{2}\right)$ Gradient $= \frac{\text{change in } y \text{ coordinates}}{\text{change in } x \text{ coordinates}} = \frac{\text{rise}}{\text{run}}$

Straight line graphs
 	
(see page 45)
 	
Parallel and perpendicular lines
 	
(see page 46)
	Straight lines
	● The **equation of a straight line** can be written in the form $y = mx + c$ where m is the gradient and c is the y-intercept
	● Two lines with gradients m_1 and m_2 are
	  ● Parallel when $m_1 = m_2$
	  ● Perpendicular (at right angles to each other) when $m_1 m_2 = -1$ or $m_2 = \frac{-1}{m_1}$

Differentiation
 	
(see page 54)
 	
Applications of differentiation
 	
(see page 56)
	Differentiation
	● The rule for differentiating powers of x is:
	$y = ax^n$ gives $\frac{dy}{dx} = anx^{n-1}$
	● When you differentiate a **constant** (a number), $\frac{dy}{dx} = 0$
	● At a stationary point the gradient is 0, so $\frac{dy}{dx} = 0$
	● For **displacement**, s, where s is a function of time
	  ● **Velocity**, v, is the rate at which displacement changes, so $v = \frac{ds}{dt}$
	  ● **Acceleration**, a, is the rate at which velocity changes, so $a = \frac{dv}{dt}$

Shape, space and measure	
Area and perimeter (see page 61) **Further area** (see page 62)	**Area and perimeter** ● The perimeter of a shape is the distance around the shape ● Area of a rectangle (or square) = length × width = lw ● Area of a triangle = $\frac{1}{2}$ × base × perpendicular height ● Area of a parallelogram = base × perpendicular height
Surface area (see page 63) **Volume** (see page 64)	**Cuboids** ● Surface area of a cuboid $= 2hl + 2lw + 2hw$ ● Volume of a cuboid $= lwh$

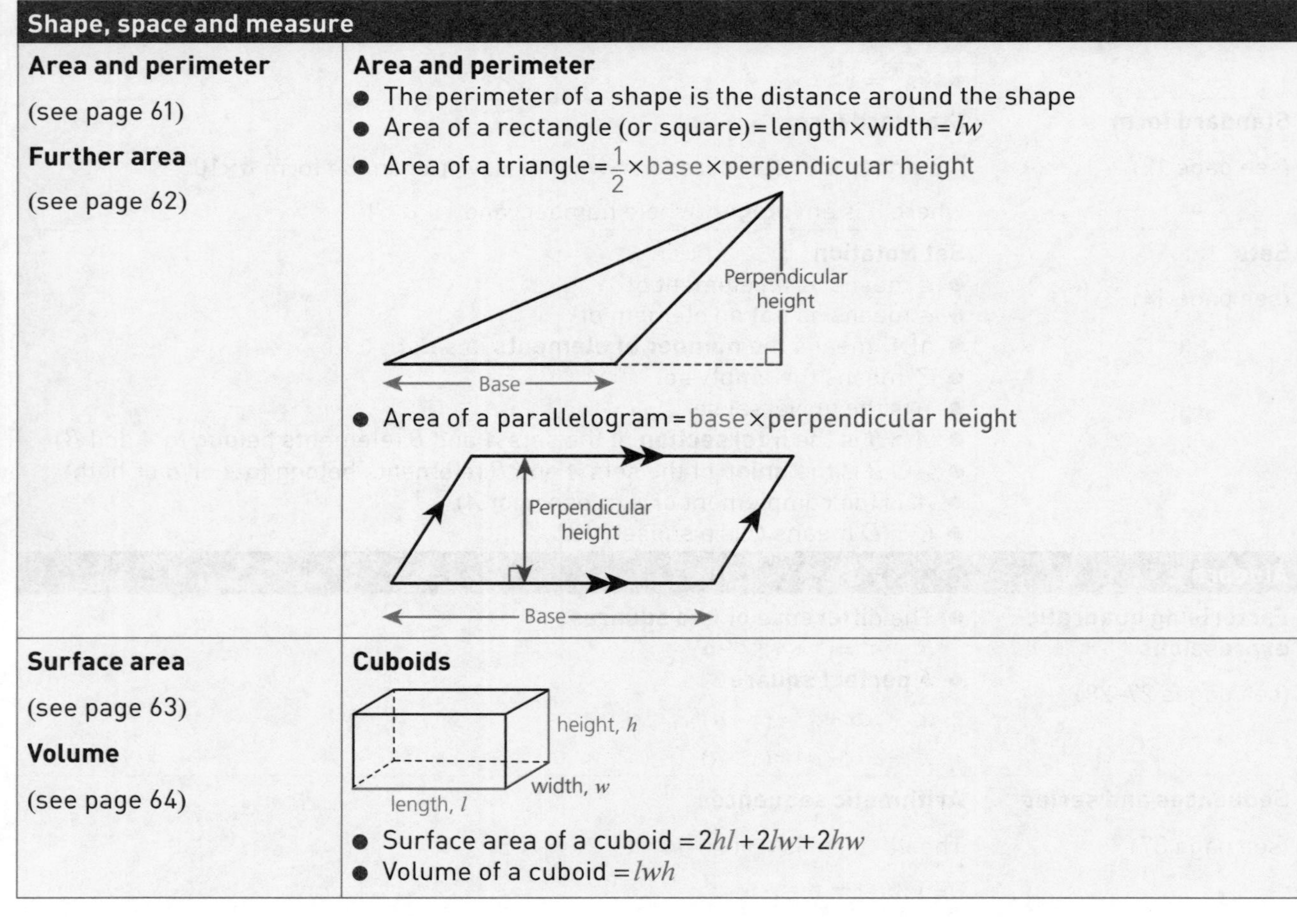

Shape, space and measure	
Circles (see page 65)	**Circles** 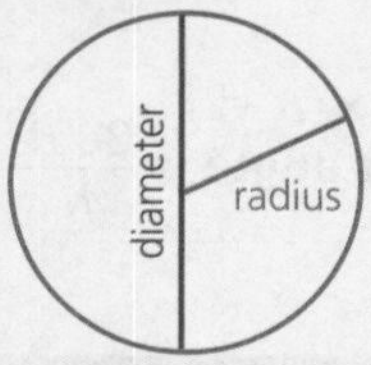 diameter, radius ● Circumference, $C = \pi d$ or $C = 2\pi r$ where d= diameter, r= radius and π =3.141 592... ● Area of a circle = πr^2
Circles (see page 65)	**Sectors** 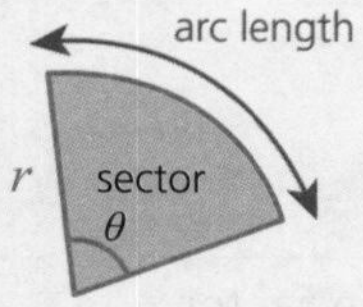 arc length, r, sector, θ ● Area of a sector of angle θ is $\frac{\theta}{360} \times \pi r^2$ where r= radius ● Arc length= $\frac{\theta}{360} \times 2\pi r$
Pythagoras' theorem (see page 70)	**Pythagoras' theorem** 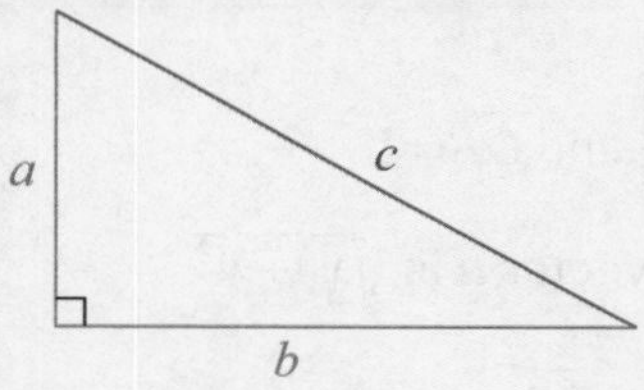 a, b, c For right-angled triangles $c^2 = a^2 + b^2$
Trigonometry (see page 72)	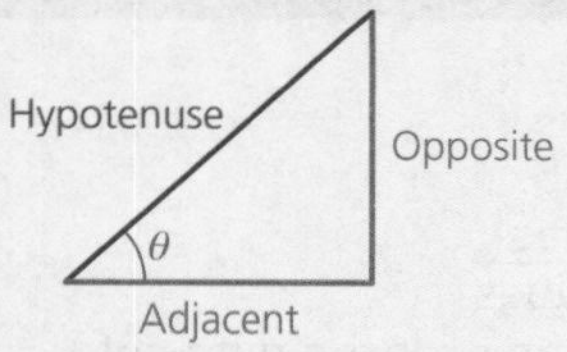 Hypotenuse, Opposite, Adjacent, θ For any **right-angled** triangle: $\sin\theta = \frac{Opposite}{Hypotenuse} = \frac{O}{H}$ $\cos\theta = \frac{Adjacent}{Hypotenuse} = \frac{A}{H}$ $\tan\theta = \frac{Opposite}{Adjacent} = \frac{O}{A}$
Compound measures (see page 80)	**Compound measures** Average speed = $\frac{\text{total distance}}{\text{total time}}$, common units are km/h or m/s Density = $\frac{\text{mass}}{\text{volume}}$, common units are g/cm^3 or kg/m^3

Make sure you know these formulae for your exam

Geometry and transformations	
Angles in polygons (see page 89)	**Angles in polygons** For an n-sided polygon: ● Sum of the **interior** angles is $180° \times (n-2)$ ● For a regular polygon the **exterior angle** is $\frac{360°}{n}$
Transformations (see page 96)	Transformations of curves (see table below)
Vectors (see page 98) **Problem solving with vectors** (see page 100)	**Vectors** For vectors $\overrightarrow{OA} = \mathbf{a} = \begin{pmatrix} x \\ y \end{pmatrix}$ and $\overrightarrow{OB} = \mathbf{b}$ ● The **magnitude** of the vector $\mathbf{a}$ is $\sqrt{x^2 + y^2}$ ● $\overrightarrow{AB} = \mathbf{b} - \mathbf{a}$ ● The resultant of $\overrightarrow{OA}$ and $\overrightarrow{OB}$ is $\mathbf{a} + \mathbf{b}$
Statistics and probability	
Averages and measures of spread (see page 109)	**Statistics** **Averages and range** ● **Mean** = $\frac{\text{total of data values}}{\text{number of data values}}$ ● **Median** = middle value of an ordered data set ● **Mode** = most common data value ● **Range** = largest data value − smallest data value
Histograms (see page 112)	**Histograms** ● Frequency density = $\frac{\text{Frequency}}{\text{Class width}}$
Probability (see pages 116–119)	**Probability** Probability of an event = $\frac{\text{number of outcomes}}{\text{total number of possible outcomes}}$ If two results A and B are **mutually exclusive** then P(A or B) = P(A) + P(B). If two results A and B are **independent** then P(A and B) = P(A) × P(B). P(at least 1) = 1 – P(none)

Transformations of curves

Graph of...	Transformation of $y = \mathrm{f}(x)$
$y = \mathrm{f}(x) + a$	Translation by $\begin{pmatrix} 0 \\ a \end{pmatrix}$ So move a units vertically
$y = \mathrm{f}(x + a)$	Translation by $\begin{pmatrix} -a \\ 0 \end{pmatrix}$ So move a units left
$y = a\mathrm{f}(x)$	**Vertical stretch**, scale factor a
$y = \mathrm{f}(ax)$	**Horizontal stretch**, scale factor $\frac{1}{a}$
$y = -\mathrm{f}(x)$	**Reflection in x-axis**
$y = \mathrm{f}(-x)$	**Reflection in y-axis**

You are given these formulae in the exam

Make sure you are familiar with the formula sheet you are given in the exam.

Topic Page	Formula
Algebra	
The quadratic formula (see page 30)	**The quadratic formula** The solutions of $ax^2+bx+c=0$ where $a\neq 0$ are given by $x=\dfrac{-b\pm\sqrt{b^2-4ac}}{2a}$
Sequences and series (see page 37)	**Arithmetic series** The sum of the first n terms $S_n=\frac{n}{2}[2a+(n-1)d]$
Shape, space and measure	
Further area (see page 62)	**Area of a trapezium** Area of trapezium $=\frac{1}{2}(a+b)h$
The sine rule (see page 75) **The cosine rule** (see page 76) **Area of a triangle** (see page 77)	**Trigonometry** In any triangle ABC **Sine rule:** $\dfrac{a}{\sin A}=\dfrac{b}{\sin B}=\dfrac{c}{\sin C}$ **Cosine rule:** $a^2=b^2+c^2-2bc\cos A$ **Area of triangle:** $\frac{1}{2}ab\sin C$
Volume (see page 64) **Cylinders, spheres and cones** (see page 66)	**Volume and surface area** **Prism** Volume of a prism = area of cross-section × **length**

Shape, space and measure	
	Cylinder 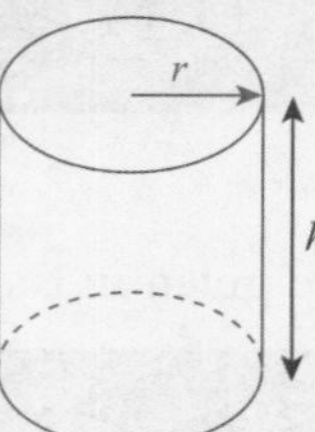 Volume of a cylinder $= \pi r^2 h$ Curved surface area of a cylinder $= 2\pi rh$
	Sphere 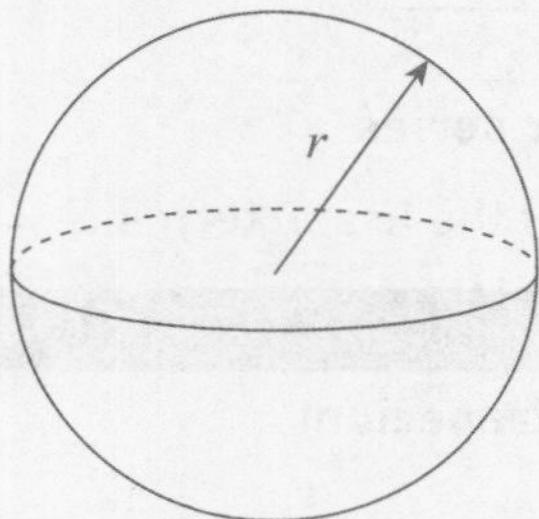 Volume of a sphere $= \frac{4}{3}\pi r^3$ Surface area of a sphere $= 4\pi r^2$
	Cone 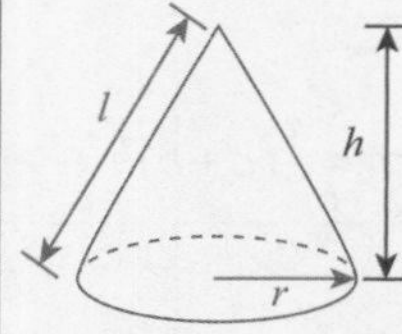 Volume of a cone $= \frac{1}{3}\pi r^2 h$ Curved surface area of a cone $= \pi rl$
Compound measures (see page 80) The formula for Pressure is not on your formulae sheet, but if you need to use it in a question it will be given to you.	**Compound measures** Pressure $= \frac{\text{force}}{\text{area}}$, common units are N/m^2 or Pascals (Pa)

During your exam

Be prepared – have the right equipment

REVISED

You need a …

- blue or black pen (and a spare pen!)
- calculator
- ruler graduated in centimetres and millimetres
- protractor
- compasses
- pencil
- eraser

Tracing paper may be used in your exam – make sure you ask for some to check any questions on transformations.

Calculators

Make sure you know how to use your calculator – it is important you are really familiar with your own calculator in the exam. Make sure you can

- use the fraction button
- use brackets
- square and square root numbers
- cube and cube root numbers
- find a power of a number, e.g. 2^{-5}
- find any root of a number, e.g. $\sqrt[4]{81}$
- enter numbers in standard form, e.g. 3.2×10^8
- use sin, cos and tan keys
- use $\sin^{-1}$, $\cos^{-1}$ and $\tan^{-1}$ keys

Check your calculator is in 'degrees' or 'D' mode:

Does your calculator give you an answer of $\frac{1}{2}$ or 0.5 when you enter $\cos 60°$?

✓ If it does, you are in the right mode!

✗ If it doesn't, then you are in the wrong calculator mode.

Ask your teacher to help you if you don't know how to change it.

Language used in the exam

REVISED

Phrase	What it means
• *Simplify ...* • *Give your answer in its simplest form ...*	Cancel down any fractions or ratios. Simplify any algebraic expressions by combining like terms.
• *Write down* • *Write* • *State*	You shouldn't need to do much (or any) working to reach the right answer. The marks will just be for giving the correct answer.
• *Work out* • *Calculate* • *Find*	You will need to do some working out or calculations to reach the right answer – write it all down! Make sure you show all your working – you can score marks for the correct method even if you make a small mistake. *Remember: no working and the wrong answer = no marks!*
• *Show clear algebraic working.* • *Show your working clearly.* • *Show that ...* • *Show by a vector method ...*	You must be very careful to show **every step** in your working! Don't skip steps, even if they seem obvious to you! If you are told to use a particular method then you must use that method otherwise you may lose **all** the marks!
• *Hence ...*	You **must** use your previous answer to help you work out this answer.
• *Hence, or otherwise, ...*	Using your previous answer will help you answer this part but you may use a different method if you prefer.
• *On the grid, draw ...*	Draw the graph or diagram as neatly as you can.
• *Use the/your graph to find ...*	Read values from the graph as accurately as you can. Add lines to the graph to help you.
• *Use ruler and compasses to construct ...* • *Draw accurately ...*	You must show **all** your construction lines as the examiner needs to see evidence that you have constructed your diagram properly. Take extra care with any measurements.
• *Sketch ...*	An accurate diagram is not needed, you don't need to use a ruler but may prefer to.
• *Give a reason for each stage in your working.* • *Give a reason for your answer.*	This is usually an angles question. Make sure you give a formal reason for each step in your working eg. 'angles in at triangle have a sum of 180°' or 'alternate angles are equal'
• *Describe fully ...*	This is usually a transformations question make sure that for ... • a **translation** ... you give a **vector** • a **reflection** ... you give **the equation of the line of symmetry** • a **rotation** ... you give the **angle, direction and centre of rotation** • an **enlargement** ... you give **the scale factor and centre.**
• *Give your answer correct to ...*	Give your answer to the degree of accuracy stated in the question – for example 3 significant figures or 1 decimal place. You may lose a mark if you don't! If you see this when solving a quadratic equation, then it is a hint to use the quadratic formula.
• *Find the exact ...*	Give your answer as fraction, surd (square root) or in terms of π and not as a decimal.
• *Diagram NOT accurately drawn*	Do not measure any sides or angles on the diagram as they won't be accurate.
• *Diagram is accurately drawn*	Take measurements (angles and/or lengths) from the diagram to help you.

During your exam...

REVISED

Watch out for these common mistakes:

✗ Miscopying your own work or misreading/miscopying the question
✗ **Using** a ruler to draw curves – draw curves free-hand
✗ **Not using** a ruler to draw straight lines
✗ Not simplifying your answer
✗ Not showing construction lines – don't rub these out!
✗ Not giving full reasons in angles questions
✗ Not describing transformations fully e.g. not stating centre, direction and angle for a rotation
✗ Rounding errors – don't round until you reach your final answer
✗ Rounding answers that should be **exact**
✗ Giving answers to the wrong degree of accuracy - use 3 s.f. unless the questions says otherwise
✗ Not showing any or not showing enough working – especially in 'show that' questions.

Check your answer is...

✓ to the correct accuracy
✓ in the right form
✓ complete...have you answered the whole question?

If you do get stuck ...

Keep calm and don't panic.

✓ **Reread the question** ... have you skipped over a key piece of information that would help? Highlight any numbers or key words.
✓ **Draw** ... a sketch can help you see the way forward.
✓ **Check the formula sheet** ... is there something there that can help you?
✓ **Look** for how you can re-enter the equation. Not being able to answer **part i)** doesn't mean you won't be able to do **part ii)**. Remember the last part of a question is not necessarily harder.
✓ **Move on** ... move onto the next question or part question. Don't waste time being stuck for ages on one question, especially if it is only worth one or two marks.
✓ **Return later** in the exam to the question you are stuck on – you'll be surprised how often inspiration will strike!
✓ **Think positive**! You are well prepared, believe in yourself!

Good Luck!

Answers

Here you will find the answers to the 'Target your revision' exercises, 'Exam-style questions' and 'Review questions' in the book. Full worked solutions for all of these (including 'show' questions that do not have short answers in the book) are available online at www.hoddereducation.co.uk/MRNEdexIGCSEMaths. Unless otherwise stated, or exact, all answers are given correct to 3 significant figures and any angles are correct to 1 decimal place.

SECTION 1

Target your revision (page 1)

1 Go online for the full worked solution

2 Go online for the full worked solution

3 Go online for the full worked solution

4 a 60 m

  b 1 : 1600

5 307 rupees

6 a 7.14%

  b £529.20

  c £8800

7 $4202.50

8 $80

9 a 2^9

  b $9a^2b$

10 $2\times3^2\times5\times7$

11 LCM = 4200 and HCF = 30

12 a i 7 920 000

   ii 0.0081

  b 9.38×10^7, 9.5×10^7, 6.1×10^8

  c i 8.9×10^7

   ii 3.87×10^{-4}

13 a i 1.308×10^{14} km³

   ii 5.8×10^{12} km³

  b 58

14 a i 2.37

   ii 0.0204

   iii 6.00

  b 56.3

15 i 3.885 m

  ii 3.895 m

16 a i $P\cap Q=\{6, 12\}$

   ii $Q\cup R=\{1, 3, 5, 6, 7, 9, 11, 12, 13, 15\}$

  b P is only even numbers and R is only odd numbers so the sets have no members in common.

  c 3, 9 or 15

  d $S=\{3, 9, 15\}$

17 a 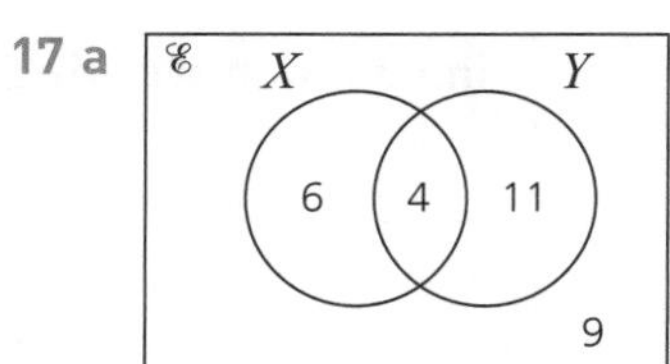

  b i 20

   ii 20

18 i $7\sqrt{7}$

  ii $9\sqrt{6}$

  iii 16

  iv $2+3\sqrt{2}$

Adding and subtracting fractions (page 3)

a–c Go online for the full worked solution

Multiplying and dividing fractions (page 4)

a–c Go online for the full worked solution

Converting decimals to fractions (page 5)

Go online for the full worked solution

Ratio (page 6)

1 70 games

2 240 cm²

3 28.8 cm

Proportion (page 7)

1 a 1.875 kg

  b 9 people

2 10 hours 15 minutes

Percentages (page 8)

1 $260.37

2 a 23%

  b 8.61 million

Further percentages (page 9)

1 a £4243.60

  b £315

2 7000

Powers and roots (page 10)

1 a 5^4
  b 5^5
  c 5^3
2 a $6x^5y^3$
  b $4xy^3$
  c x^2

Factors, multiples and primes (page 11)

1 $2^2 \times 3^2 \times 5 \times 7$
2 a 84
  b 23 520

Standard form (page 12)

1 a i 291 000
    ii 0.000 56
  b 3.7×10^{-8}, 3.74×10^{-8}, 2.9×10^{-7}
  c i 2.84×10^5
    ii 2.84×10^{-2}
    iii 2.84×10^7
    iv 2.84×10^{-4}
2 1833
3 8×10^{3x}

Degree of accuracy (page 13)

1 i 26.049 421 86
  ii 26.0
2 i 3.09 to 3 s.f.
  ii 0.571 to 3 s.f.

Sets (page 14)

1 a i $A \cap B = \{3, 6, 9, 18\}$
    ii $B \cup C = \{1, 2, 3, 4, 6, 9, 16, 18\}$
    iii $A' \cap C = \{1, 4, 16\}$
  b x is 12 or 15
  c $D = \{4, 16\}$
2 i 21
  ii 3
  iii 20
  iv 8
  v 5

Surds (page 16)

1 $-2-6\sqrt{5}$; $a=-2$ and $b=-6$
2 $13\sqrt{3}$; $n=13$
3 Go online for the full worked solution

Review questions: Number (page 17)

1 a Go online for the full worked solution
  b 36
2 Go online for the full worked solution
3 $n=3$
4 $x=350$
5 $2.5 \times 10^{4n-1}$
6 i 23
  ii 42
  iii 62
7 $a=6$ and $b=-3$

SECTION 2

Target your revision (page 18)

1 a $6a^2+12a$
  b $8a^3b-2a^2b^2$
  c $7c+9$
  d $2d^2+11d+12$
  e $9e^2-12e+4$
  f f^3+2f^2-5f-6
2 a $a(a+4)$
  b $3b(2-b)$
  c $4c(3b-5c)$
  d $6(2d-1)$
  e $12e^3f(3e^2-2f)$
3 a $x=-12$
  b $x=\frac{5}{4}$
  c $x=3.5$
4 a $x \geqslant -2$

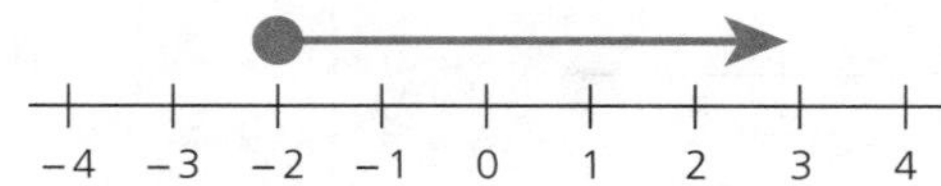

  b $x > 3$

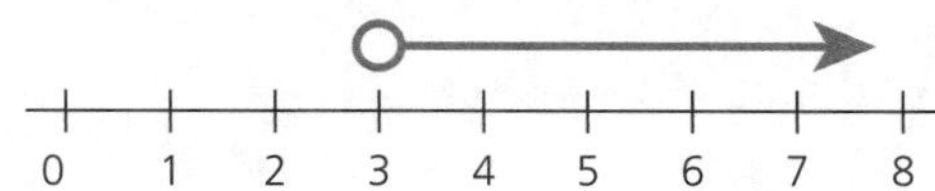

  c $-4 \leqslant x < 2$

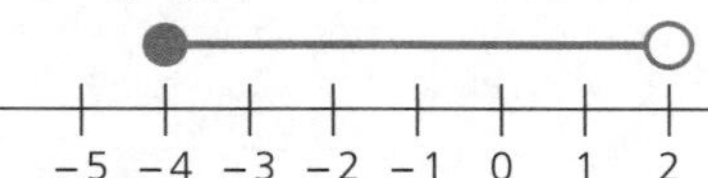

5 0, 1, 2 and 3
6 a $x=-2$ and $y=3$
  b $x=7$ and $y=-2$
7 $T=4x+12y$
8 a i $R=4$
    ii $s=4$
  b i $C=9$
    ii $a=31$
9 a $w=\frac{(x+6)^2}{4}$
  b $x=\frac{y+3}{y-1}$
  c $y=\frac{x}{\sqrt{1-2x}}$

10 i $(4x+5)(4x-5)$
ii $(x+4)(x-5)$
iii $(3x-1)(x-5)$

11 i $x=4$
ii $x=\frac{1}{3}$ or $x=2$

12 $x=-0.436$ or $x=3.44$

13 i $x=3.5$
ii $x=3$

14 i $2(x-5)^2-6$
ii $x=5\pm\sqrt{3}$
iii $(5, -6)$

15 i $-4\leqslant x\leqslant 4$

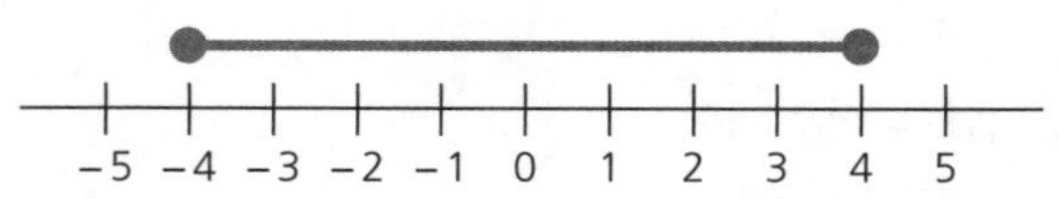

ii $x<-2$ or $x>1$

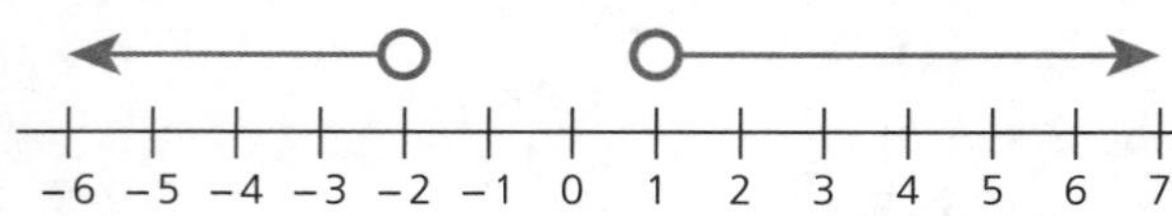

16 a $\frac{x+4}{2x+1}$
b $x=-2$

17 i 

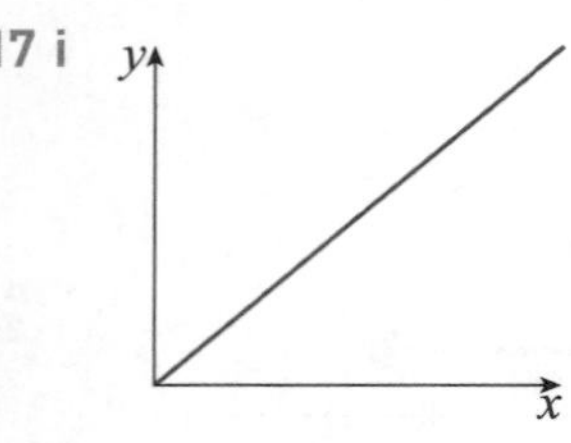

ii

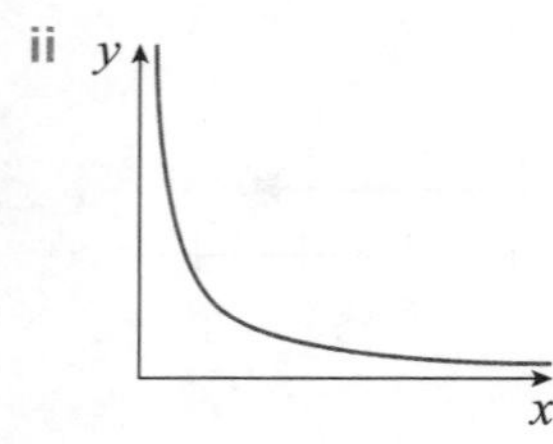

18 i $P=\frac{5}{\sqrt{q}}$
ii $q=\frac{1}{25}$

19 $x=2, y=-1$

20 a $20-2n$
b $d=5$

21 a i 9
ii 1
b $x=3$

22 a $x=-1$
b i 7
ii $f^{-1}: x\mapsto \frac{8-x}{x}$

23 a and b go online for the full worked solution

Expressions (page 20)

1 a $6x-10x^2$
b $12x^3y+18x^2y^2$

2 a $7x+20$
b $8x^2+2x-3$

3 x^3+3x^2-6x-8

Factorising (page 21)

1 a $3d(5c+6e)$
b $5y(4y-1)$

2 $6st(3t-6s+5)$

3 $4c^3d^2(2c^3+3d^2)$

4 $4(x-5)(2x-7)$

Solving linear equations (page 22)

1 a $x=-5$ b $x=\frac{1}{2}$

2 a $x=\frac{1}{3}$ b $x=\frac{3}{4}$

Solving inequalities (page 23)

1 $x>-2$

2 i $-\frac{3}{2}\leqslant x<4$
ii $x=-1, 0, 1, 2$ or 3

3 i $x\leqslant -4$
ii

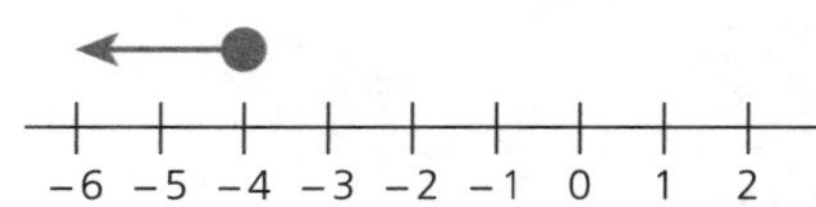

Simultaneous equations (page 24)

1 $a=3, b=\frac{1}{2}$

2 $c=-3, d=4$

3 $e=2, f=-\frac{1}{2}$

Formulae (page 25)

1 $A=25x+100y$

2 i $P=220$
ii $x=3.5$

Rearranging formulae (page 26)

1 $r=\sqrt{\frac{100\pi-A}{4\pi}}$ 2 $b=\frac{1-3a}{2-5a}$

3 $s=\frac{3t+5}{2-4t}$ 4 $y=\frac{x^2}{1-2x^2}$

Factorising quadratic expressions (1) (page 27)

1 $(8x+1)(8x-1)$

2 $(x+7)(x-8)$

3 $(x-9)(x-5)$

4 $(x+6)^2$

Factorising quadratic expressions (2) (page 28)

1 $(2x+5)(x+4)$

2 $(3x+2)(x-5)$

3 $(2x+5)^2$

4 $2(3x+2)(2x-1)$

Solving quadratic equations (page 29)

1 i $x=\frac{1}{2}$ or $x=-\frac{1}{2}$

ii $x=0$ or $x=\frac{9}{4}$

2 i $x=\frac{1}{2}$ or $x=-3$

ii $x=\frac{3}{2}$

3 i $x=-3$ or $x=1$

ii $y=1$ or $y=5$

The quadratic formula (page 30)

1 $x=-5.70$ or $x=0.702$

2 $x=-1.64$ or $x=2.14$

3 i $p=-14$

ii $r=-1.61$ or $r=3.11$

Setting up equations (page 31)

a 17 cm

b i Go online for full worked solution

ii 44.8 cm

Completing the square (page 32)

1 i $(x+4)^2-5$

ii $x=-4\pm\sqrt{5}$

iii -5

2 i $3(x-4)^2-9$

ii

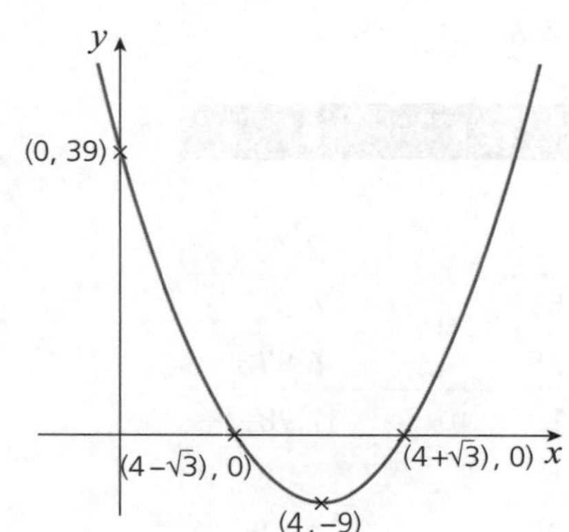

Solving quadratic inequalities (page 33)

1 $x\leqslant 0$ or $x\geqslant 5$

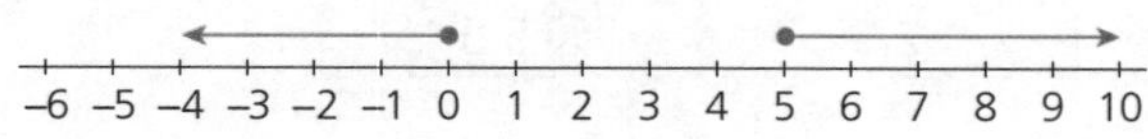

2 $x<-6$ or $x>3$

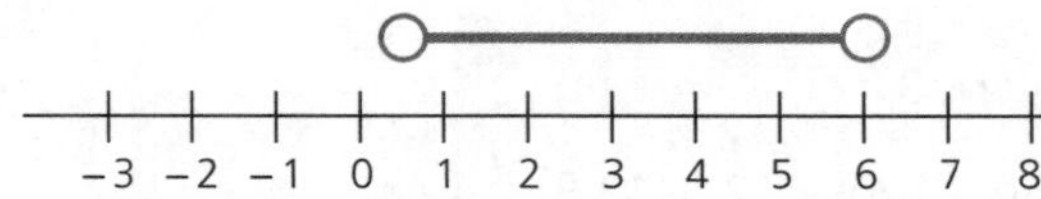

3 $\frac{1}{2}<x<6$

−3 −2 −1 0 1 2 3 4 5 6 7 8

Algebraic fractions (page 34)

1 $\frac{3(x-1)}{(x+3)(x-3)}$

2 $\frac{2(x-1)}{x}$

3 $\frac{x-2}{2x-3}$

4 $x=\pm 2$

Proportion (page 35)

1 a $M=4n^3$

b $M=32$

c $n=\frac{1}{2}$

2 a

R

A

b $R=\frac{24}{5A}$

Non-linear simultaneous equations (page 36)

1 $x=-2$, $y=2$

2 $x=\frac{3}{2}$, $y=9$

3 $x=0$, $y=3$ or $x=-3$, $y=0$

Sequences and series (page 37)

1 a $107-7n$ b -6390

2 35 terms

Functions (page 38)

i $x<1$

ii 1

iii $\mathrm{gf}(x)=\frac{6}{x+1}$

iv $x=2$

Inverse functions (page 39)

i $x\neq 5$, so the excluded value is $x=5$

ii 3

iii $\mathrm{f}^{-1}(x)=\frac{5x}{x-2}$

iv $x=0$ or $x=7$

Answers

Reasoning and proof (page 40)

1–3 Go online for the full worked solutions

Review questions: Algebra (page 41)

1 Go online for the full worked solution

2 i $\frac{1}{5} < t < 3$

ii $-5(t-1.6)^2 + 12.8$; 12.8 m, 1.6 s

3 $5\sqrt{6}$

4 a Go online for the full worked solution

b 44 cm

5 a 5

b Go online for the full worked solution

c $x = 4.63$ or $x = 10.4$

d $x > 6$

6 a $(3n+1)$ km

b 650 km

SECTION 3

Target your revision (page 42)

1 (0, 3)

2 a $D(0, -3)$

b $\text{Gradient}_{AB} = \text{Gradient}_{CD} = 1$

$\text{Gradient}_{BC} = \text{Gradient}_{CD} = -\frac{2}{3}$

c $ABCD$ is a parallelogram.

3 a

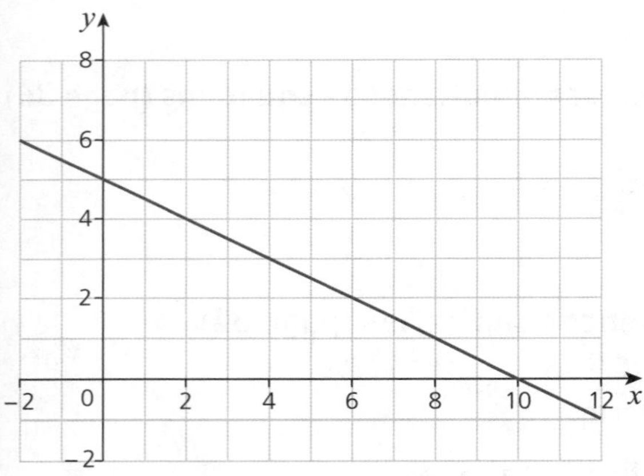

b $-\frac{1}{2}$; (0, 5)

4 a $\frac{2}{3}$; (0, −4)

b $y = 2x - 13$

5 $y = 4x + 11$

6 $4x + 2y = 3$

7 a

x	−2	−1	0	1	2	3	4
y	13	3	−3	−5	−3	3	13

b

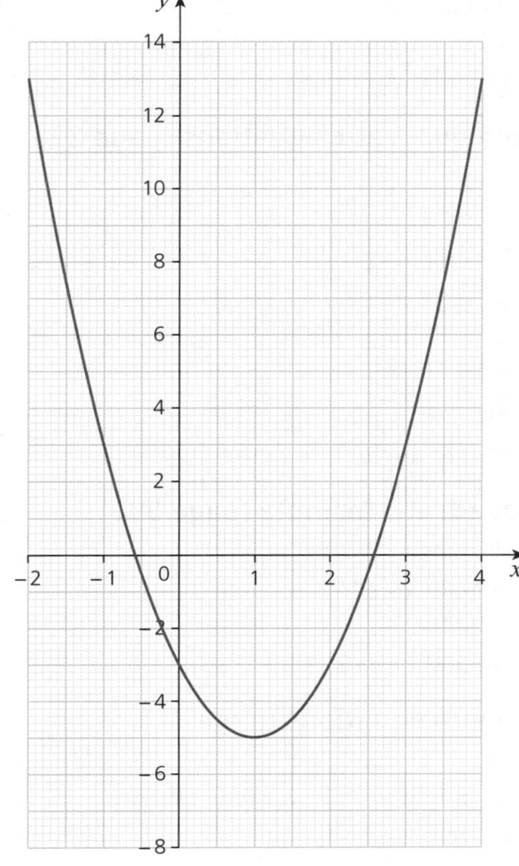

c Any value from 7.8–8.2

d $x = -1.2$ or $x = 3.2$

8 i $x = -1.5, y = -7.9$
or $x = 1.7, y = -1.6$ (all to 1 d.p.)

ii $x = -0.4$ or $x = 1.2$ (all to 1 d.p.)

9 $y \geqslant x$, $x + y \geqslant 6$ and $y < 5$

10 a

x	y
−2	4.5
−1	−2
−0.5	−4.5
−0.25	−6.875
−0.1	−12.98
0.1	7.02
0.25	1.125
0.5	−0.5
1	0
2	5.5

b

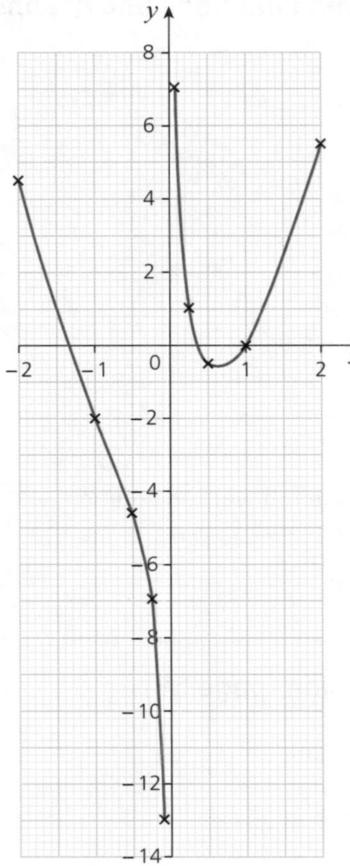

c $a = -0.5$, $b = 4.5$

11 a 7 m/s

b 2.8 seconds

c −2.7 m/s^2

12 a 53

b (−1, 13)

13 (1, 3)

14 a Go online for the full worked solution

b 1728 m^2

Coordinates (page 44)

a $A(-4, 1)$

b $x = 4$

c $D(-2, -3)$

Straight line graphs (page 45)

1 $y = -2x + 5$

2 i

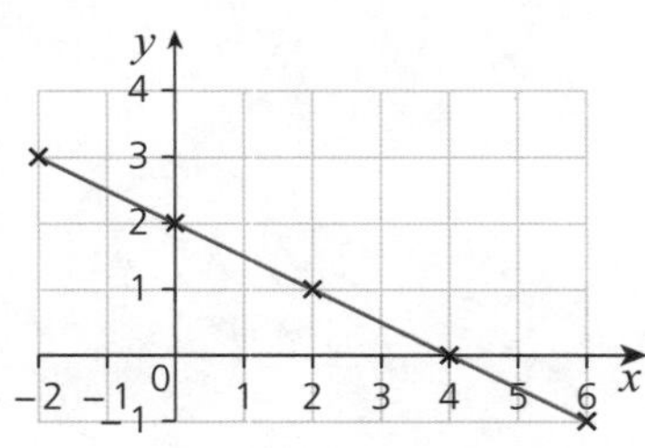

ii $-\frac{1}{2}$; (0, 2)

Parallel and perpendicular lines (page 46)

1 a $k = -2$ b $y = \frac{4}{3}x - 6$

2 a Go online for the full worked solution

b $y = \frac{1}{3}x$

Quadratic curves (page 47)

a

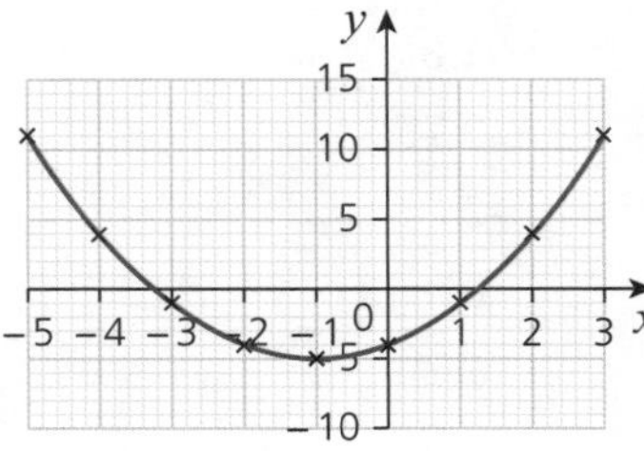

b Any value from 3.5–4.5

c $x = -3.2$ or $x = 1.2$ to 1 d.p.

Graphical solution of equations (page 48)

a

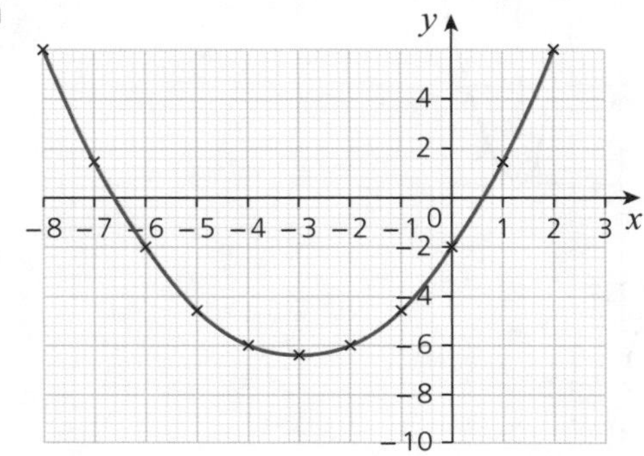

b i $x = -3.6, y = -6.3$

or $x = 1.6, y = 4.3$ (to 1 d.p.)

ii $x = -4.4$ or $x = 0.4$ (to 1 d.p.)

Inequalities and regions (page 49)

i

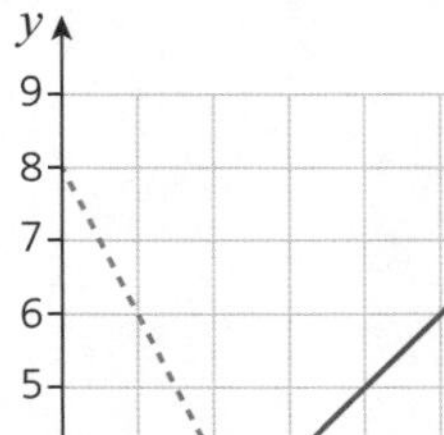
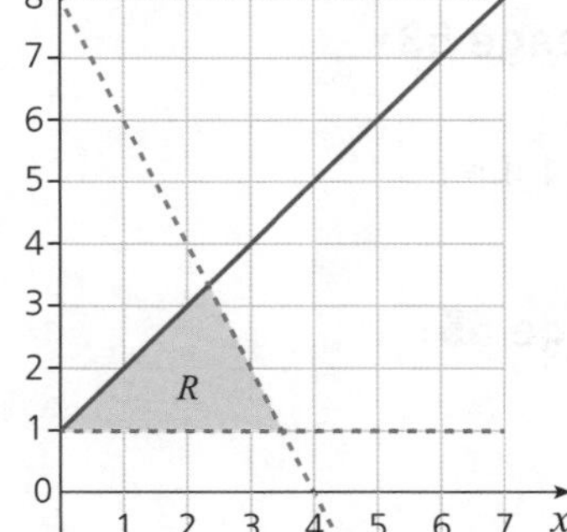

ii (1, 2), (2, 2) or (2, 3)

Further graphs (page 51)

The answers are estimates so allow ±0.1

1 a i −1.9

ii $x = -1.3$ and $x = 1.2$

b $-1.3 < k < 1$

2 a

x	y
−3	3.3
−2	−2
−1	−6
−0.5	−8.8
−0.3	−11.6
0.3	1.8
0.5	0.75
1	−2
2	0
3	4.7

b

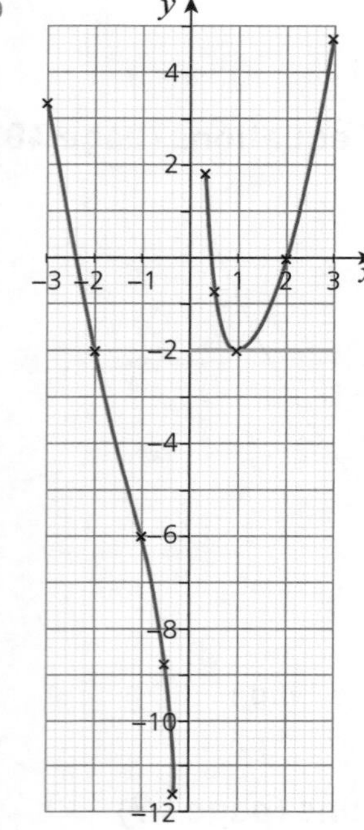

c $k=1$

d −4

Real-life graphs (page 53)

a 1.5 hours

b 7.7 km/h (correct to 1 d.p.)

c 1.3 km

Differentiation (page 55)

1 $\left(\frac{1}{2}, \frac{5}{2}\right)$

2 a (1, 3)

b 2.25

Applications of differentiation (page 56)

a 49.5 m

b $v=24t-\frac{3}{t^2}$

c 72 m/s^2

Review questions: Functions and graphs (page 57)

1 a $x+y=-1$

b $p=3$; $q=3$

2 $\frac{45}{4}$ square units

3 a $b=-9$ and $c=5$

b $y=3-7x$

4 a i −1.3 m

ii 0.4 s, 4 s and 8 s

iii 2.3 m

b $a=2$, $b=6.3$

c −1 m/s

5 a Go online for the full worked solution

b 4 cm, 6 cm and 12 cm; 288 cm^2

SECTION 4

Target your revision (page 59)

1 20.4 cm^2

2 a 0.0325 m^2

b 650 000 mm^2

3 i 36 cm^2

ii 56.7 cm^2

4 6 cm

5 a 0.123 m^3

b 315 mm^3

6 252 cm^3

7 81.4 m^2; 35.2 m

8 12.9 cm^2; 14.6 cm

9 a 480π cm^3

b 312π cm^2

10 382 cm^3

11 a i 3.8 cm

ii 6 cm

b 7.11 cm^2

12 $a=5$

13 a 15.1 cm

b 38.7°

14 109 m

15 a 64.1°

b 15.9 cm

16 48.3 cm^2

17 A(0°, 1)

B(90°, 0)

C(180°, −1)

18 774 g

Area and perimeter (page 61)

a 43.5 cm^2

b i 0.00435 m^2

ii 4350 mm^2

Further area (page 62)

1 41 cm^2

2 $x = \frac{5}{3}$

Surface area (page 63)

$x = 4.90$

Volume (page 64)

540 cm^3

Circles (page 65)

1 Area = 57.4 cm^2; perimeter = 30.6 cm

2 $\left(\frac{25}{3}\pi + 12\right)$ cm

Cylinders, spheres and cones (page 67)

1 a 314 cm^3

b 283 cm^2

2 a 3.75 cm

b Go online for the full worked solution

3 3810 cm^3

Similarity (page 69)

1 a 18.9 cm

b 6.4 cm

c 53.4 cm^2

2 32 cm

Pythagoras' theorem (page 71)

1 $5\sqrt{2}$ cm

2 13.9 cm

Trigonometry (page 73)

1 107.4°

2 6.93 cm

3D trigonometry (page 74)

25.1°

The sine rule (page 75)

1 31.1°

2 7.01 cm

The cosine rule (page 76)

1 $CAB = 118.8°$, $ABC = 23.9°$, $ACB = 37.4°$ (all to 1 d.p.)

2 12.0 cm

The area of a triangle (page 77)

1 43.3 cm^2

2 91.0 cm^2

The graphs of trigonometric functions (page 79)

1 a

x	$y = \tan x°$
0	0
30	0.6
45	1
60	1.7
75	3.7
90	–
105	–3.7
120	–1.7
135	–1
150	–0.6
180	0

b

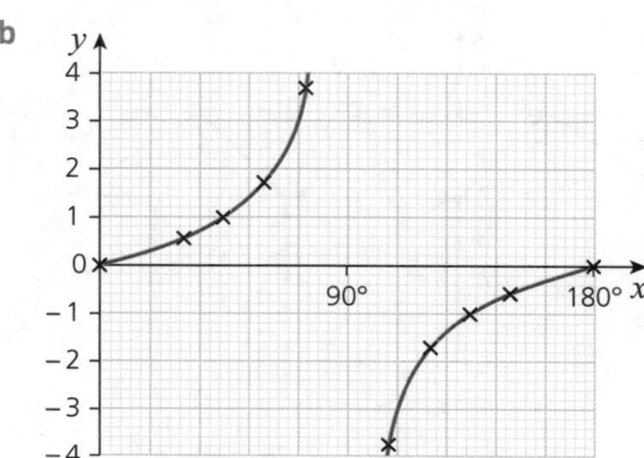

c 63° (accept within 3° of this value)

2 a $A(-90°, -1)$ $B(90°, 1)$ $C(180°, 0)$

b i $\frac{1}{2}$ ii $-\frac{1}{2}$

Compound measures (page 80)

1 694 km/h

2 686 g

Review questions: Shape, space and measure (page 81)

1 24 cm

2 $\frac{5}{32}$

3 131.2°

4 a 4.59 m b 14.4 m

5 67.7 cm^2

6 a 317 kg b $\frac{27}{50}\pi$ m^2

SECTION 5

Target your revision (page 83)

1 a 69° b 42° c 138°

2 256°

3 101°

4 a 1 b 1

5 100°

6 117°

7

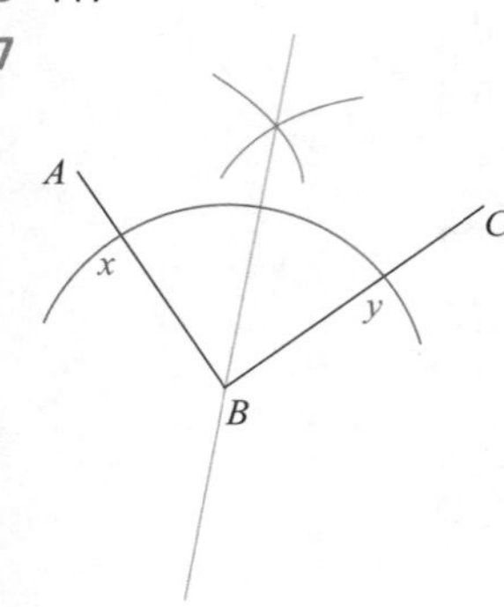

8 A translation of $\begin{pmatrix} 9 \\ -1 \end{pmatrix}$

9

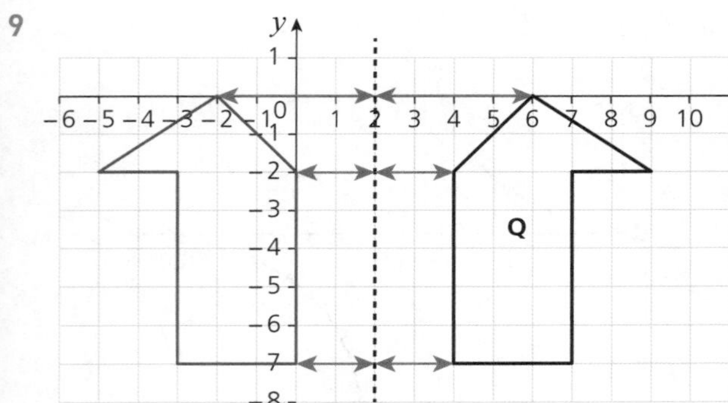

10

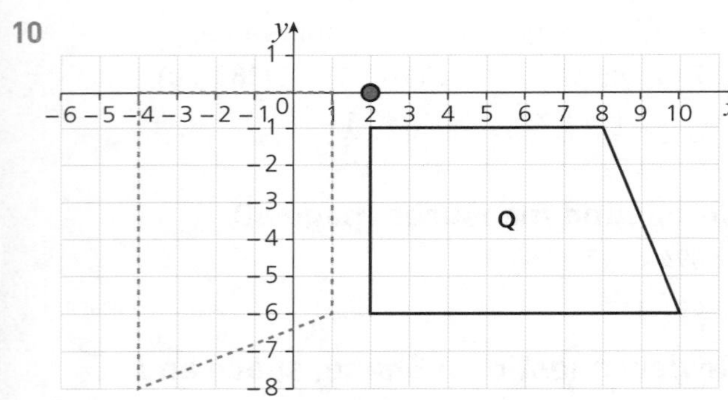

11 Enlargement, scale factor 2, centre (0, 2)

12 Graph **G** $f(x-8)$ or $y=(x-5)^2$

13 a $\begin{pmatrix} 6 \\ -2 \end{pmatrix}$ b $\begin{pmatrix} 8 \\ 46 \end{pmatrix}$ c $\sqrt{29}$

14 a $2\mathbf{a} + \frac{3}{2}\mathbf{b}$ b $\mathbf{b} - \frac{8}{3}\mathbf{a}$

2D shapes (page 85)

a B

b D

c i A ii C

Angles (page 87)

1 48°

2 $x = 35°$, alternate angles equal

$y = 103°$, angles in triangle add up to 180°

3 $x = 107°$

4 a $x = 25$ b Angle C

Bearings (page 88)

1 075°

2

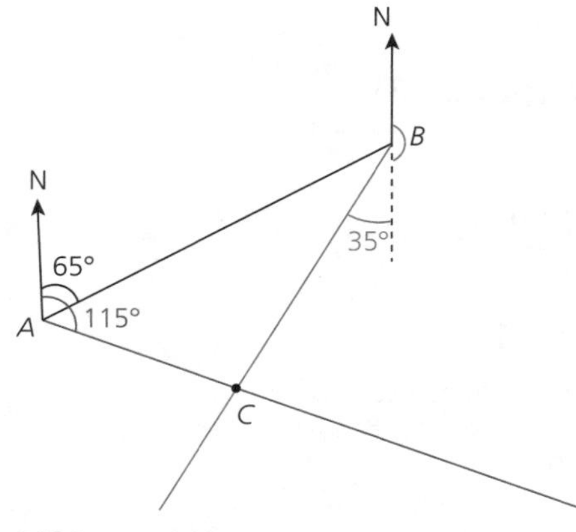

3 019°

Angles in polygons (page 89)

1 a 24 b 3960°

2 $n = 6$

Constructions (page 91)

1 a

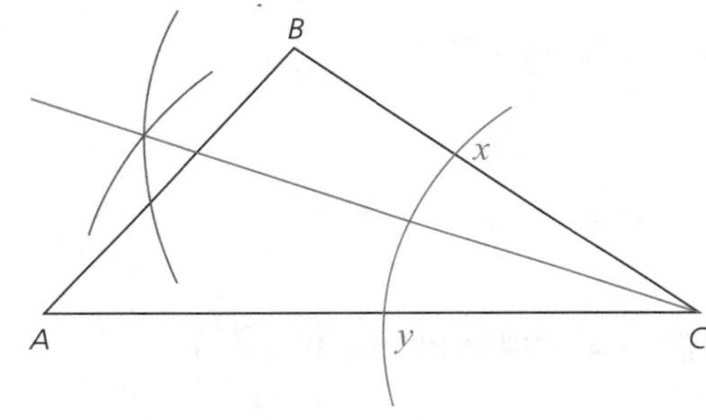

b

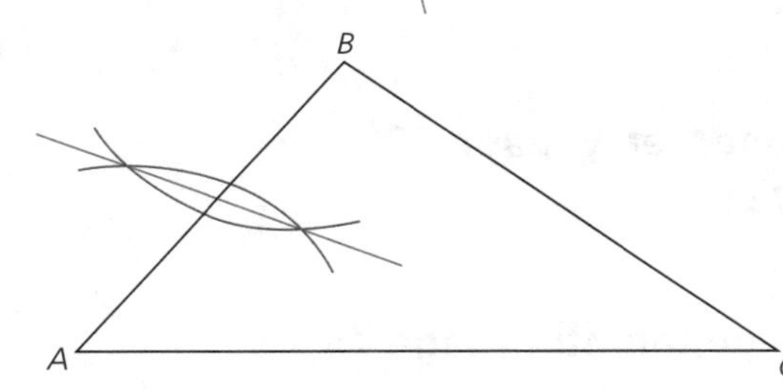

2

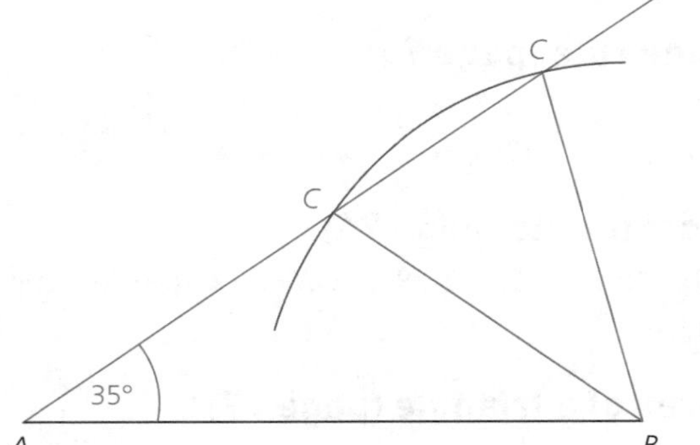

3 Go online for the full worked solution

Translations (page 92)

1

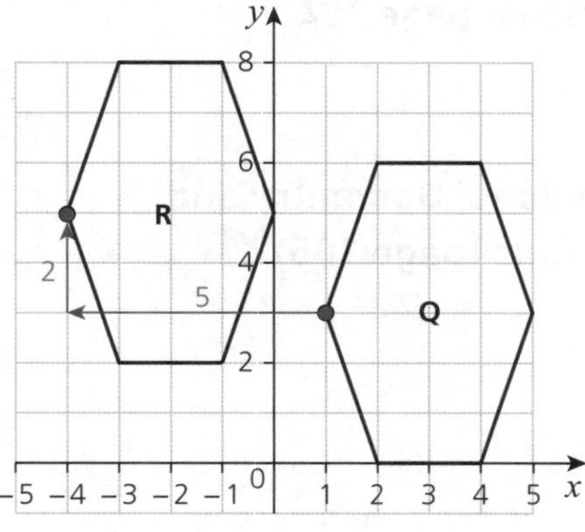

2 Translation of $\begin{pmatrix} 4 \\ -1 \end{pmatrix}$

Reflections (page 93)

1

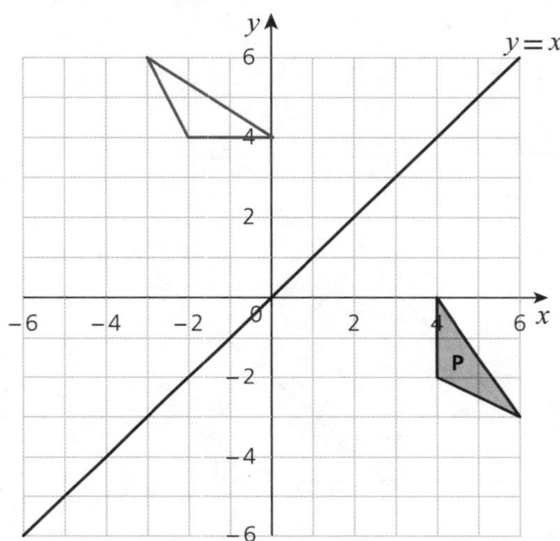

2 a $y = 2$

b

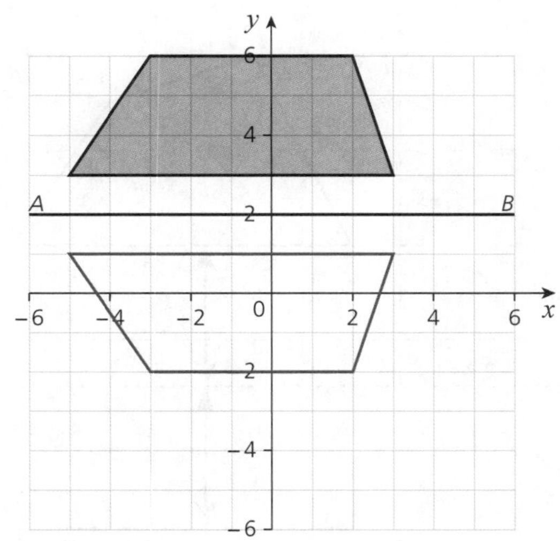

Rotations (page 94)

1 Rotation, 90° anticlockwise, centre (1, 0)

2

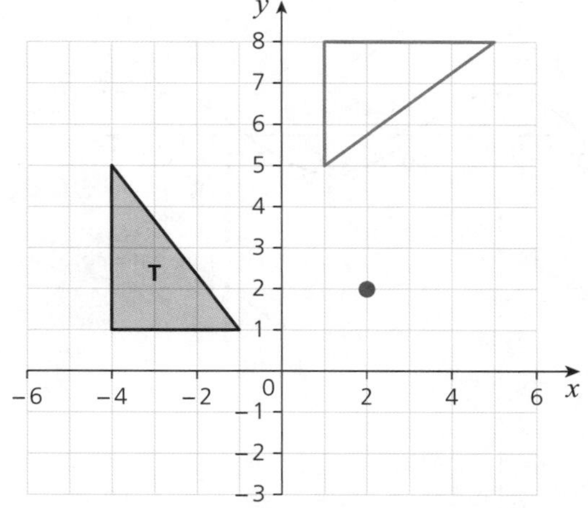

Enlargements (page 95)

1 Enlargement, scale factor 2, centre (0, 0)

2

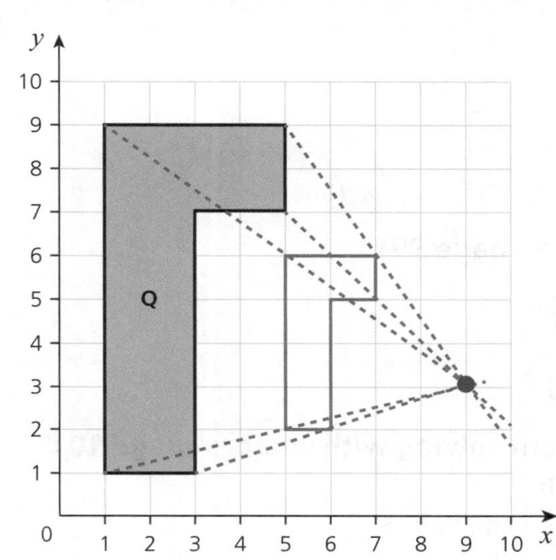

Graphs and transformations (page 97)

1 a

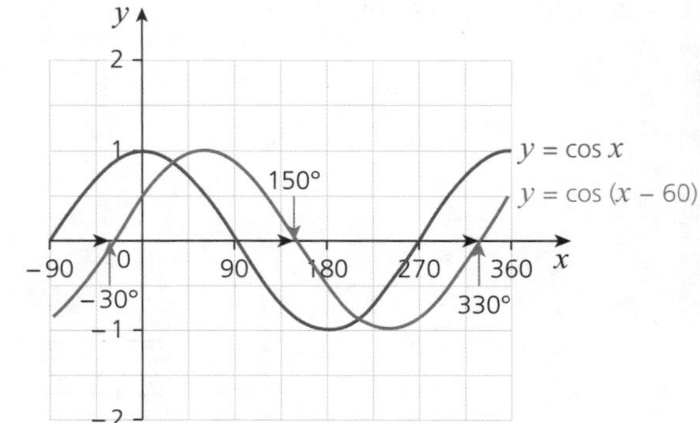

b

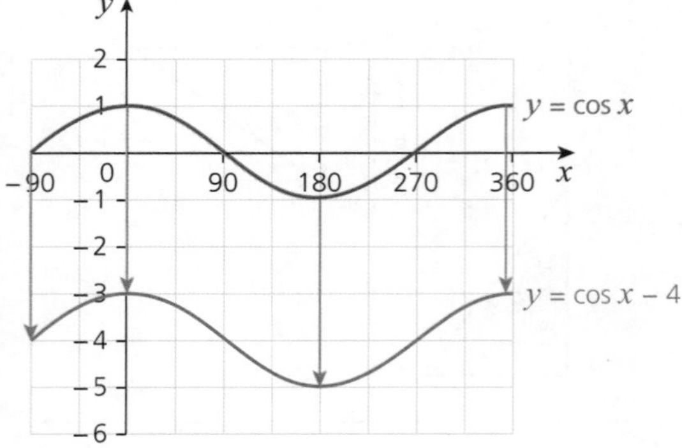

c

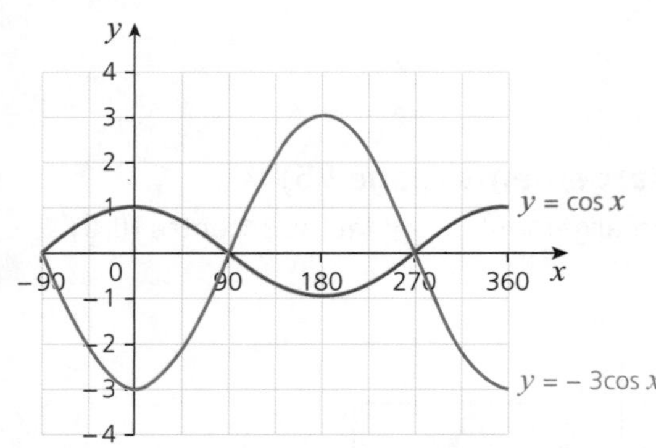

2 a (−1, −1) b (3, 4)

Vectors (page 99)

1 $\begin{pmatrix} 6 \\ 22 \end{pmatrix}$

2 $\sqrt{170}$

Problem solving with vectors (page 101)

1 $3\mathbf{a}-\mathbf{b}$

2 a $3\mathbf{a}+6\mathbf{b}+9\mathbf{c}$

b $2\mathbf{a}+4\mathbf{b}+6\mathbf{c}$

c $\overrightarrow{NO}$ and $\overrightarrow{MX}$ are not multiples of each other so are not parallel.

3 $18\mathbf{b}+30\mathbf{a}$

Circle theorems (page 104)

1 73°

2 a 80° b 65°

Review questions: Geometry and transformations (page 105)

1 a 32° b 148°

c 32° d 74°

2 140°

3 107° (to 3 s.f.)

4 a 81°, alternate segment theorem

b 44° as angle at the centre is twice the angle at the circumference

5 a

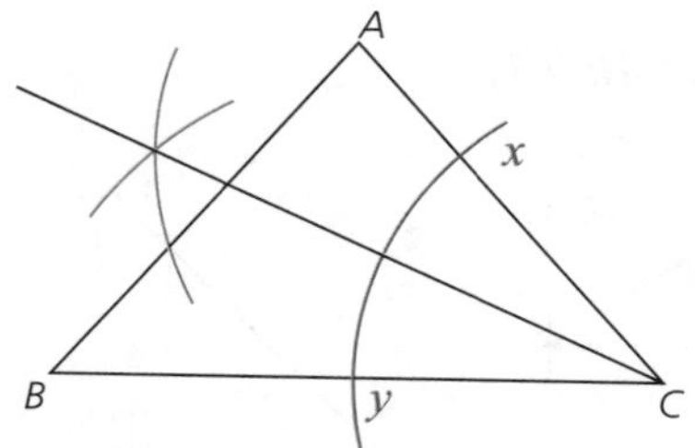

b

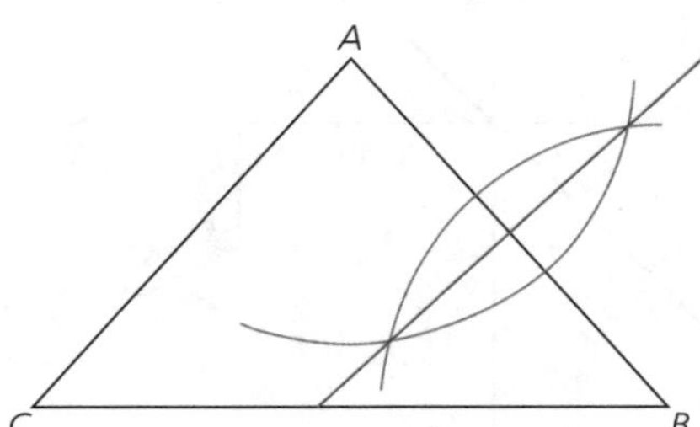

6 a

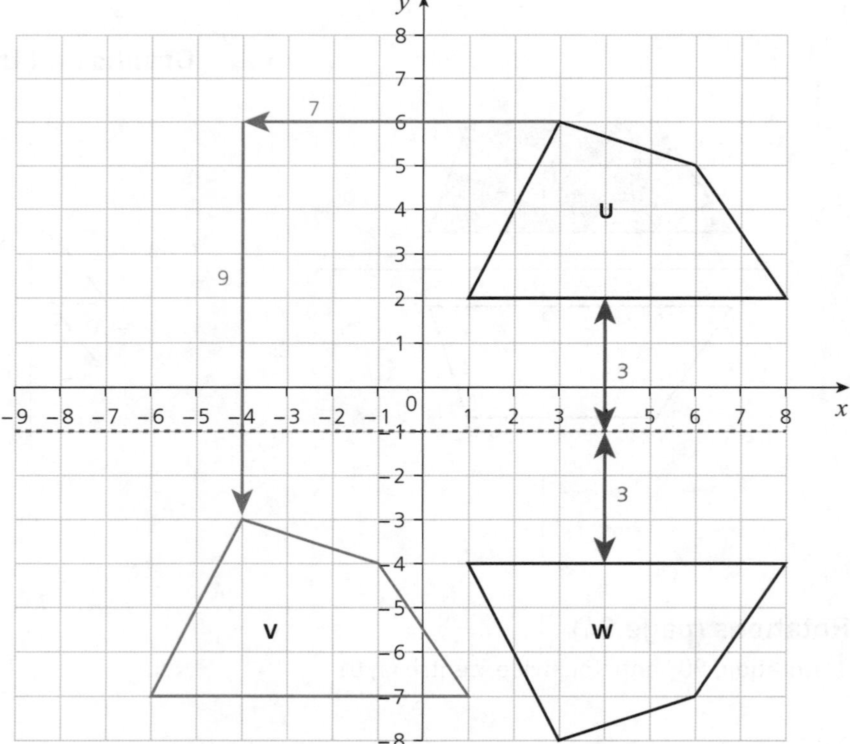

b Reflection in the line $y=-1$

7 $a=3$, $b=2$, $c=-2$

8 a $\overrightarrow{CA}$ and $\overrightarrow{CX}$ are scalar multiples of each other and have C as a common point. So CXA is a straight line.

b $\overrightarrow{YX}$ is a scalar multiple of $\overrightarrow{CD}$ so these vectors are parallel.

SECTION 6

Target your revision (page 107)

(page 107)

1 a 31 b 24.25 c 24.5

2 12

3 a $0<d\leqslant 10$ b 14.5 c $10<d\leqslant 20$

4 17.8 years

5 9

6 a

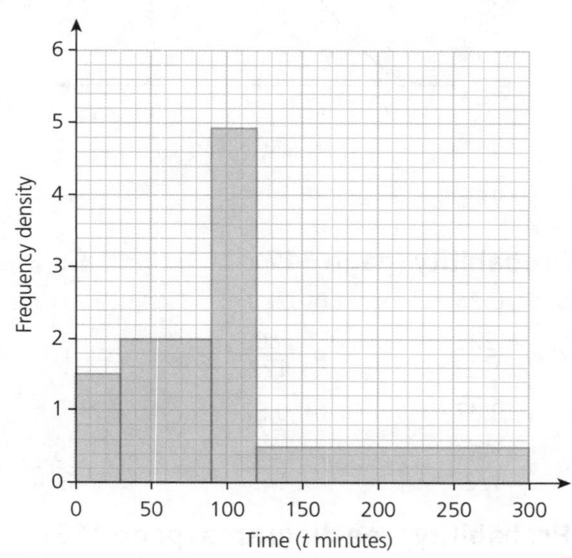

b 188

7 a

Frequency	Cumulative frequency
10	10
15	10 + 15 = 25
26	25 + 26 = 51
66	51 + 66 = 117
32	117 + 32 = 149

b

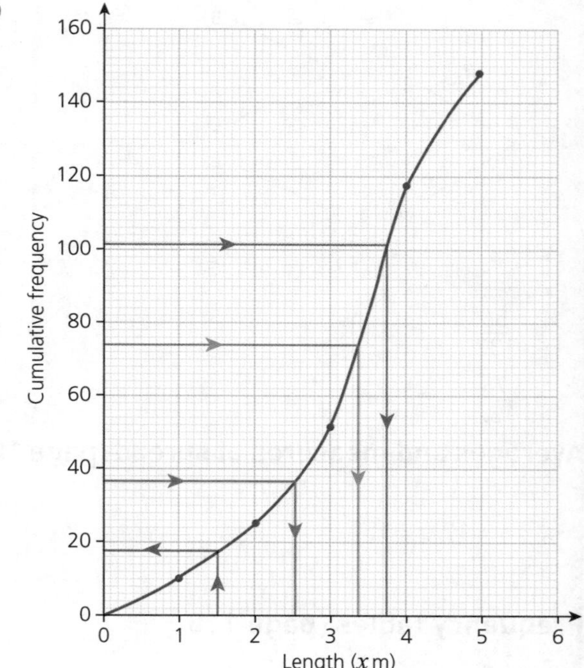

c 3.38 m

d 1.2 m

e 130

8 a 16 minutes

b 19 and 12

c 7

d Peter

9 i x at 1

ii x at $\frac{1}{4}$

iii x at 0

10 15

11 a 0.25 b 0.75

12 0.85

13 i $\frac{9}{36}$

ii $\frac{18}{36}$

14 a 0.12

b 0.18

c 0.42

15 a

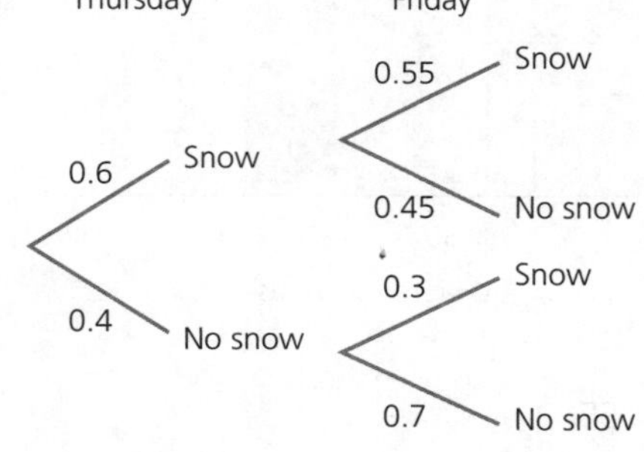

b 0.39

16 a

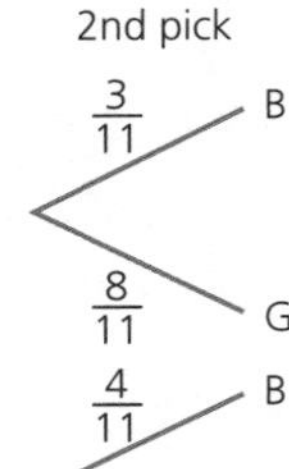

b $\frac{56}{132}$

c $\frac{76}{132}$

Averages and measures of spread (page 109)

1 27

2 $x = y = 12, z = 15$

3 6

Frequency tables (page 110)

1 a 2.4

b 2

c $\frac{21}{60}$

2 a 4 b 3.21 c 2 d 4

Grouped frequency tables (page 111)

a $3 < L \leqslant 4$

b 5

c 2.8

Histograms (page 113)

1 a 9.76%

b 15

2 a

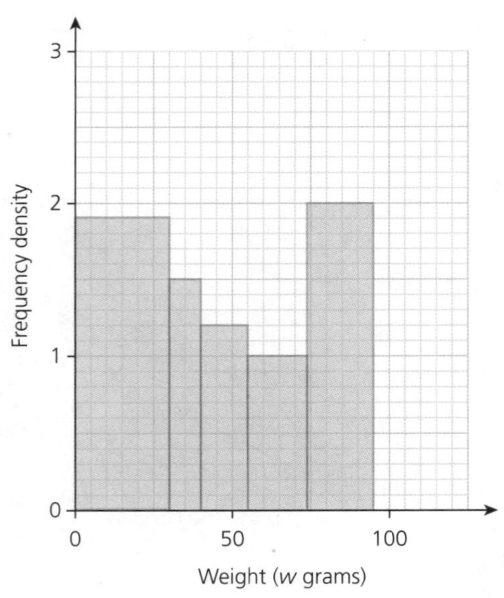

b 59.6%

c $40 < w \leqslant 55$

d 84

Cumulative frequency (page 115)

a

Weight (w grams)	Cumulative frequency
$0 < w \leqslant 5$	4
$0 < w \leqslant 10$	32
$0 < w \leqslant 15$	62
$0 < w \leqslant 20$	81
$0 < w \leqslant 25$	85

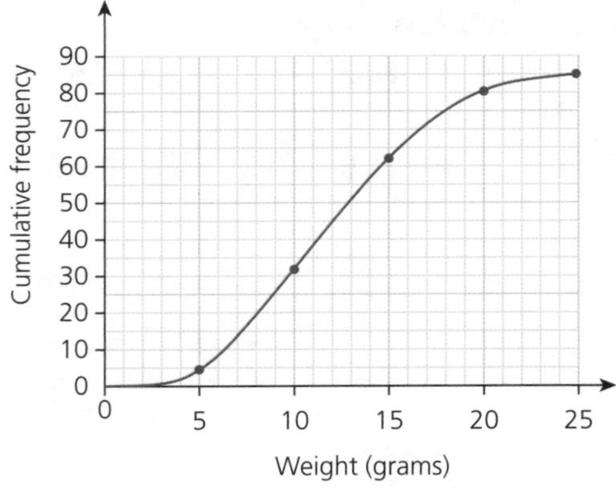

b 96.5%

Probability (page 117)

1 a 0.26 b 220

2 a $\frac{225}{1225}$ b $\frac{600}{1225}$ c 1

3 a 0.07 b 0.52

4 a $\frac{55}{122}$ b $\frac{41}{145}$ c $\frac{55}{320}$

Probability tree diagrams (page 118)

1 a

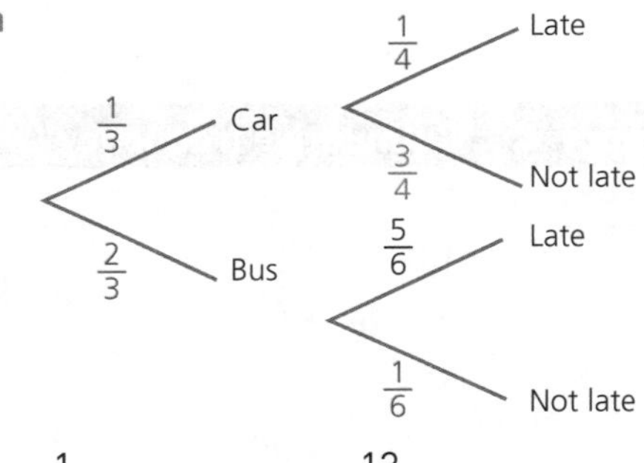

b $\frac{1}{12}$ c $\frac{13}{36}$

2 a

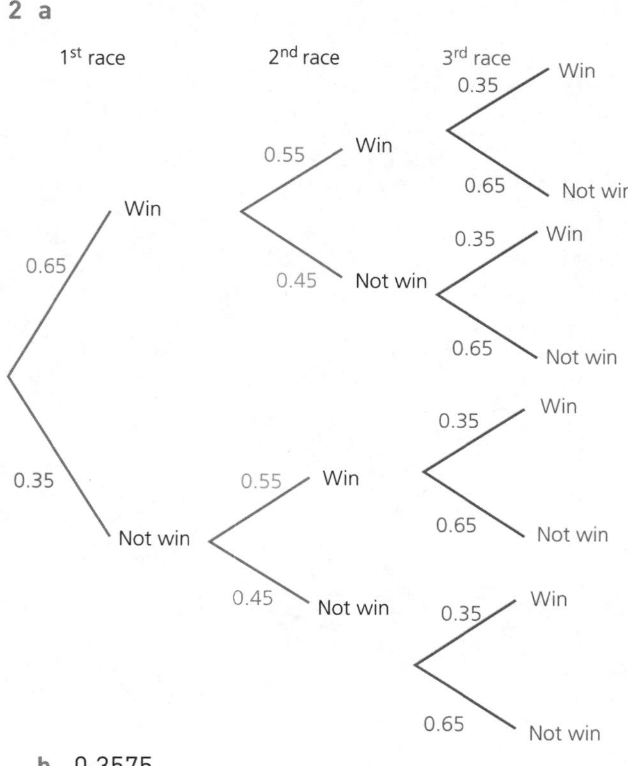

b 0.3575

c (i) 0.370

(ii) 0.898

Conditional probability (page 119)

1 a

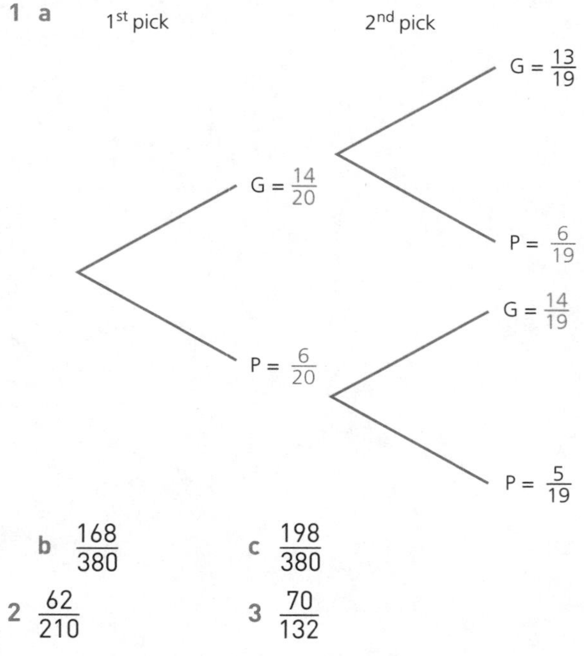

b $\frac{168}{380}$ **c** $\frac{198}{380}$

2 $\frac{62}{210}$ **3** $\frac{70}{132}$

Review questions: Statistics and probability (page 120)

1 $x = 17$, $y = 20$, $z = 33$

2 41.1 g

3 a €7192.50

b €50.30

4 a

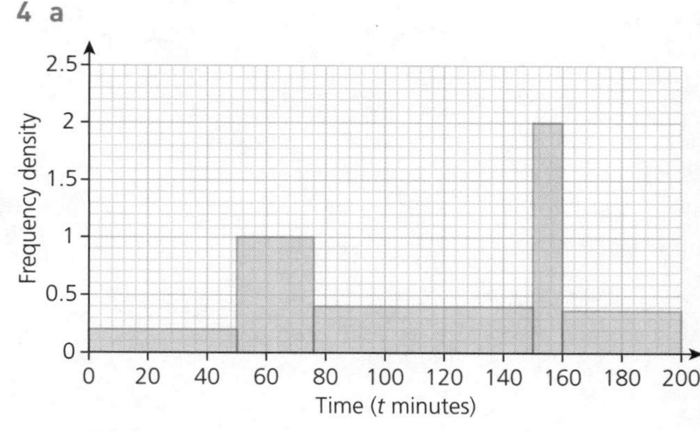

b 59

5 a

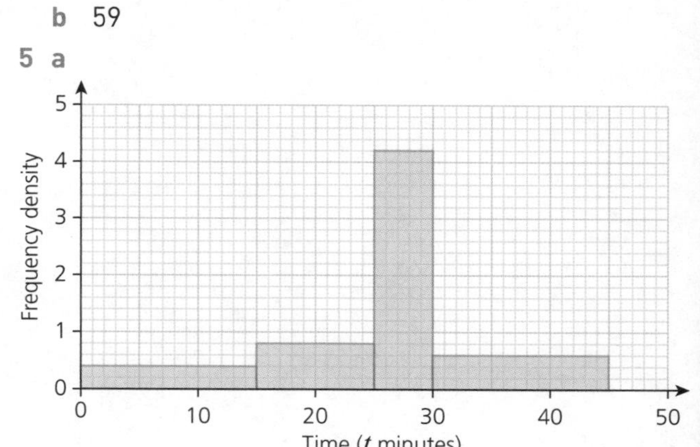

b 31

6 a

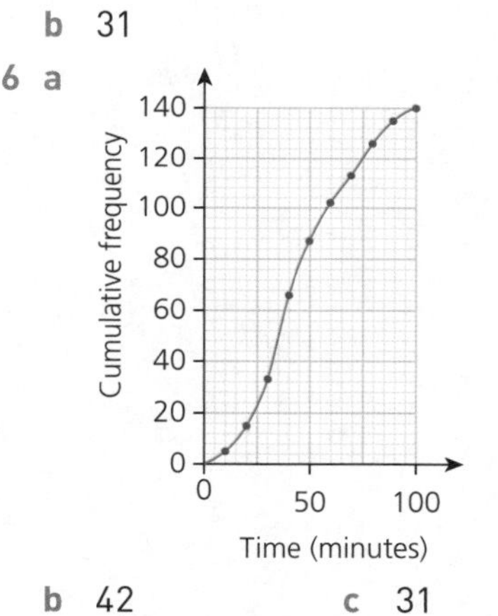

b 42 **c** 31 **d** 20

7 a

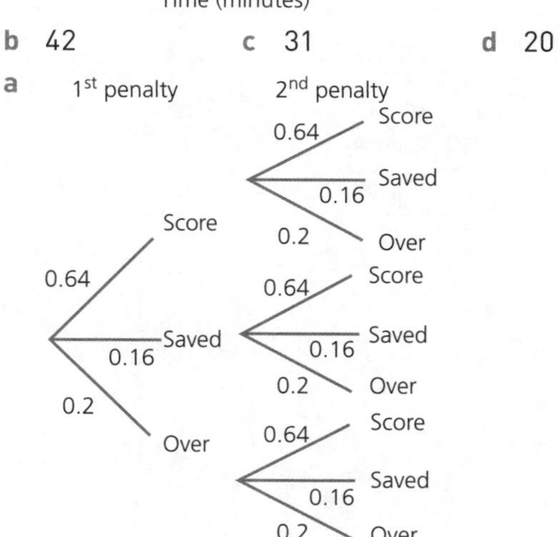

b 0.8704

8 $\frac{210}{625}$